U0107916

食 物 小 传

大米

Rice

A Global History

〔美国〕勒妮·马顿 著

王艺蒨 译

北京联合出版公司
Beijing United Publishing Co.,Ltd.

目录

巧妇难为无米之炊。

——中国谚语

不论你身在何处，都有可能每天食用米饭。事实上，如今世界上三分之二的人口都在践行这样的饮食习惯，其中多数人生活在那些拥有悠久的水稻种植历史的国家。此外，在有大量以米为主食的移民定居的国家，大米的消费量也在不断增加。中国、南亚、印度北部与从印度尼西亚到缅甸、日本的其他亚洲国家，以及西非和中非国家，都是水稻的发源地。而在其他地区，水稻则是外来作物。在早期，利润的驱使和养活劳动人口的需要（两者不一定同时）成为刺激大米种植和贸易的两大因素。人口迁移——无论主动还是被迫——和水稻的移植总是并行的。水稻和人类都适应了新的环境。

如果你从纽约去广州旅游，你的早餐大概会是稀饭或粥。每天有数以百万计的人食用这种通常由隔夜剩饭做成的米粥。在加利福尼亚的萨克拉门托，你也能喝到稀饭，那是由于因 19 世纪 50 年代的加利福尼亚淘金热而去往美国的广东人的后代定居在那里。在那时，大米作为留居加利福尼亚的四万中国劳工的日常食物而得以进口。然而，直到19世纪末加利福尼亚的水稻种植才开始作为一种产业发展起来。20世纪20年代，商业化水稻种植加速发展。1850年，刚刚成为"州"的加利福尼亚进口的大米大多来自中国。但到了1950年，萨克拉门托谷的水稻种植就已经非常成熟了。甚至到了2008年，加利福尼亚州产的大米的50%已经出口到日本、韩国、乌兹别克斯坦和土耳其了。

在美国，广东移民开了第一批起初主要面向中国食客的中餐厅。渐渐地，美国人也对这些"东方"食物产生了兴趣，于是一些中国厨师开始在私人家庭工作。对于这些广东籍的中国厨师来说，简单的一碗米饭几乎可以搭配所有菜肴（在以大米为主食的国家，人们通常都是直接吃白米饭的）。我们通常说的炒饭，则是源于家里处理剩饭的巧妙做法。现在，炒饭可以单独作为菜单上的一道菜肴，这也反映了中餐为亚洲以外的人群做出的改良。在1965年美国移民政策放松后，大量的中国移民从台湾、香港和福建涌入纽约、旧金山、洛杉矶和其他城市，拓展了"华裔"的定义。当然，地域性的米食也跟着他们一起来到这片土地。

如果你在南卡罗来纳州的查尔斯顿庆祝新年，烩菜豆米饭大概是你餐食中的一部分。这道由大米和豇豆或黑眼豆制作而成的非洲传统食物来源于西非人的拿手菜。这种饭食不仅由奴隶带到英国、荷兰、法国、西班牙、葡萄牙和尚且不叫美国的美洲殖民地（并使水稻成为那里的主要粮食作物），甘蔗、棉花、烟草、靛蓝植物等殖民地种植园的劳工也将它传播到加勒比群岛、巴西、秘鲁、古巴和墨西哥等地。此外，印度人也作为契约工人前往西印度群岛，随后中国人及其他国家的人也去往当地。起初主要为奴隶和劳工而进口的大米成为一种商品。时至今日，如果你仍在吃烩菜豆米饭，那你大概率有非洲或加勒比血统，也可能两者皆有，你可能身处美国东南部、加勒比群岛、墨西哥，也同样可能定居在底特律、密歇根或印第安纳州加里市。

你和朋友小聚，吃着寿司，喝着清酒。你会在哪里呢？东京？很有可能，但先等一下——你听到了葡萄牙语和英语——原来，你在圣保罗或者伦敦。虽然寿司在全球米制食品领域中发展得较晚，但它在城市的影响力几乎是全球性的。再来看看加州卷，这种由内而外的卷饭有时用糙米制作而成，里面卷有牛油果、黄瓜、胡萝卜、蛋卷或香草，却没有生鱼。它既可能出现在新加坡高档日本餐厅的菜单里，又可能是上海犹太餐厅提供的菜品。在东京的烹饪学校里，厨师甚至会学习如何"正确地"制作加州卷。

在飞回纽约的班机上，你在航空杂志中读到了一些很著名的米制餐食：海鲜饭、意大利烩饭、印度香饭和手抓饭。这四道饭食通常出现在家庭餐桌、餐厅、露天市场、快餐车里和节日庆典中，它们分别与西班牙、意大利、印度和伊朗有关。最近的历史研究表明，它们的起源几乎可以追溯至莫卧儿王朝和伊斯兰商人活跃时期。

刚回到纽约，机场的孟加拉裔的出租车司机就邀请你尝试了一种叫米花沙拉（jhal-muri）的小吃。将经过柠檬和香菜调味的膨化米粒和花生、洋葱碎、辣椒拌在一起，便化身为这道加尔各答特色菜肴。在纽约街道的一侧，来自南印度的街头小贩们在卖印度薄饼（dosas），一种塞满米粒和小扁豆的可丽饼（crêpes）；而在另一侧，巴基斯坦人在卖着鸡肉饭、咖喱饭和印度香饭。

1947 年，英国对印度的殖民统治结束。然而，无论是在印度独立前、过程中还是独立后，都有印度移民持续涌入英国，使两国在饮食文化上的联系绵延了一个世纪有余。许多英国殖民者从印度回来后仍然怀念印度美食的风味，一些娶了印度夫人或者将当地厨师带回国的英国人，则得以满足食欲。印度移民将他们的烹饪传统融入新的生活，也融入了我们的生活。而众多街头摊贩、餐厅以及之后的加工食品制造商和超市进一步加速了融合菜在世界范围内的传播。

你因暴饮暴食而感到不舒服。为了缓解你的肠胃不适，你准备喝一碗温和的奶油米糊（和稀饭很相似）。你还准备喝一杯欧洽塔（horchata），一种源自中美洲和墨西哥的清凉米制饮品。你开始尝试以糙米膨化饼干和玄米绿茶为主的饮食，偶尔来上一杯百威啤酒；没错，大米也是百威啤酒的主要原料。你给孩子们做了点大米脆当点心，还给你的狗狗喂了食物。等等,宠物食品里也有米吗?你猜得不错。大米无处不在,而且以上提到的不过是"冰山一角"。终于,你的身体缓过劲儿来,你又准备了一个撒满葡萄干和开心果的米布丁带去参加晚宴,作为餐后甜品。

国家与文化

无论是在蕉叶中蒸制，在瓦罐中煨炖，又或是在电饭锅中蒸煮，大米常常宛若一块白色画布，其上描绘着绚烂的饮食文化。辣白菜、酱油、咸猪肉、腊鱼、山药、牡蛎等调味品和／或米饭配菜都蕴含着传统饮食的起源。在以大米为主的饮食文化中，一碗米饭不仅是检验你是否进食正餐的标准，还是饮食中主要的热量提供者。通常情况下，人们更青睐吃精米，但糙米也被广泛食用——后者需要更少的人力加工，采购成本更低，而且相比精米具有更高的营养价值。碾磨过的米可以被制成粉条、薄饼、糕点和饼干。米粉被当作酱料、布丁、香肠、婴儿辅食和宠物食品的增稠剂，也被添加在化妆品里。其他含有大米成分的产品还有米皮、醋、味噌、米糠油、啤酒、米露和米酒。膨胀、爆破、研磨成片的米被用来制作谷物片、小零食、饼干、点心或者面包。茎秆和谷壳还能被编织成凉鞋和垫子，或者作为燃料燃烧。

大米是如何生长的

大米是一种适应力很强的谷类植物，在大多数环境下可以生长，但是产出却不甚理想。因此，灌溉水稻占栽培水稻的50%，2010年，灌溉水稻产量占约7亿吨水稻产出的75%（而其中的30%~35%会在去壳、碾磨和抛光的过程中损失掉）。湿地水稻先在苗圃育苗，然后再移植到稻田里。一年一熟、二熟到三熟（取决于水稻品种、地理位置和气候）

的水稻产出是一个艰苦且对体力要求极高的过程。水稻农业主要分布在热带地区，例如中国南方、东南亚、印度等其他亚洲国家或地区，以美国、墨西哥为主的北美洲，以及非洲、南美洲的国家或地区。雨水灌溉的低地水稻的产量占全球产出的20%，而陆地水稻仅占5%，后者又被称为旱稻，主要分布在南美洲和非洲。最后，还有一种可以生长在深50厘米甚至更深的水里的深水稻，是孟加拉国和其他洪水泛滥的河谷地区的主要农作物，水稻在上升的水中生长迅速，最高可达4米。

随着世界人口增加、城市扩大，水稻生产者也在努力跟上人们需求增长的脚步。除了提高产量和采用更环保的灌溉方式以外，立体农业或许能解决一些问题。这种方法本质上是陆地梯田水稻的一种变体，包括在建筑物侧面垂直层或巨大的温室里种植水稻。与此同时，人们正在开发新的高产水稻品种以满足日益增长的需求。他们正在研发有更佳风味和可持续性的"有机"水稻。与此同时，人们正在对大米中发现砷的最新消息进行评估，并探索减少砷的方法。本来，砷是一种通过植物根茎吸收的天然矿物质，但最近，尤其在加利福尼亚州的水稻中，砷的含量极高。虽然人们正在调查此事，但到目前为止还没有一个确切的解释。

大米的文化和食用价值

精米适合于多种烹调方法和文化传统。在1965年以前，如果你吃过"puso"（一种用叶子包裹的米饭），你大概是个菲律宾人，住在马尼拉。然而在 1965 年美国移民限制放宽后，你也有可能生活在有大量菲律宾移民定居的加利福尼亚戴利城。

糙米是一种低脂、低胆固醇、少钠，但富含八种氨基酸、B 族维生素、铁、钙和纤维素的复合碳水化合物。它营养丰富，很有嚼劲，饱含坚果香气。那为什么大多数人都更喜欢精米呢？脱去糠层和胚芽的精米更易保存，烹饪时熟得更快，价格便宜，容易消化而且令人产生满足感。除此之外，它雪白的颜色拥有着长久的象征价值：纯洁、干净、尊贵的身份和优良的品质。

利比亚的妇女正在筛米，这就是她们获得洁白无瑕的大米的方法。

一麻袋的糙米，等待交易中。

煮熟的大米既可以像粥一样柔软，也能像手抓饭底部的"tahdig"①（呈金黄色硬壳状）一样硬。米饭可咸可甜，或软或脆，冷热皆宜，甚至能作为一种工具：黏米饭能特别容易地用手指按压成小碗形状，然后将其蘸些酱汁，再裹进小块的肉、鱼或蔬菜。总之，大米拥有无限的可能。

在大米消费量最大的地区，当地居民每日摄入的热量有三分之二来自大米。有四个原因可以大致解释为什么大米有如此多的忠实粉丝：首先，它与其他美味互补。例如在手抓饭或者米布丁中，它可以同时保留自己的风味和口感。其次，米饭可以作为辛辣食物的缓冲剂，比如搭配印度咖喱、黄咖喱或者秋葵浓汤。再次，大米能锁住水分，使食物更加鲜美多汁，例如制作葡萄叶卷饭或者米肠。最后，米饭

---

① 波斯美食，类似于锅巴。——本书注释，除特别说明，均为编注

黑色背景下的红碗白米饭——美极了！

的加入通常会降低餐费。

最重要的是，大米原本是挣扎在温饱线上的人群的主要食物，却在近些年成为发达国家的高档食品。统计数据表明，随着人们越来越富裕，他们消费了更多的动物蛋白。随着可支配收入的增加，人们开始寻求更高品质的蛋白质食品。然而，在一些工业化的国家，健康问题比彰显地位更加重要，因此糙米的摄入量不断增加，同时肉类和鱼类的摄入量在减少。当然，因为世界人口数量不断增长，对于大米，特别是精米的需求量也在攀升。

根据位于菲律宾的国际水稻研究所（IRRI）的统计，2010 年下列 20 个国家的人均大米消费量（按重量计算）是最高的，以每人每年消费的千克数来衡量。虽然中国、印度和印度尼西亚的人均消费量并不在榜单的前列，但由于它们庞大的人口数量，它们仍然是世界上主要的大米生产国和消费国。

2010 年人均大米消费量（千克）

文莱	245
越南	166
老挝	163
孟加拉国	160
缅甸	157
柬埔寨	152
菲律宾	129
印度尼西亚	125
泰国	103
马达加斯加	102
斯里兰卡	97
几内亚	95
塞拉利昂	92

几内亚比绍	85
圭亚那	81
尼泊尔	78
朝鲜	77
中国	77
马来西亚	77
韩国	76

相比而言，巴西在 2008 年的人均大米摄入量是 44 千克（部分原因在于寿司消费，尤其在圣保罗），而美国的人均大米摄入量只有 11 千克。

亚洲稻（栽培稻）是稻属的一个主要品种，大多数栽培稻都是它的后代。由于不同的品种在感官属性上差异很大，风土条件——植物生长的特定环境，包括气候、地质和地形——不仅对植物遗传很重要，而且对味道、颜色和香气也很重要。世界上有超过 11.5 万个水稻品种。众所周知，水稻种植不仅需要重劳力，还要耗费大量水资源，才能确保其高产。你或许会思考，那些以大米为生的人为什么要劳心劳力地生产这种食物呢？而答案正隐藏在这个问题之中："人们忠诚于这种谷物恰恰是因为他们付出如此大的努力培育了它。"（即便当人们迁居到城市里，这种偏好也通过各种形式表现出来。例如，在城市里，吃一顿没有米饭的午餐仍然意味着没有"真正地"进餐，也许只算是吃了一种零食。）对于任何一个移民群体来说，他们都渴望吃到自己家乡的大米。并且这种田间劳动也更划得来——每英亩水稻的产量和所能提供的热量都高于小麦、玉米、大豆和小米。

大米的品种和形态

你更喜欢哪种大米呢？糙米、巴斯马蒂米、糯米还是美国产的"本大叔"呢？虽然我们比较熟悉这些叫法，但它们并不是真的这样被分类的，因为米粒大小、形状、颜色、黏性、味道和香味各不相同。脱去谷壳（也称为稃，是保护谷粒的外壳）后剩下的是糙米，它仍保有米糠、胚芽和一些营养元素。再去除这些元素，你就得到了几乎是纯淀粉的精米。

米的种类：彩虹系列。

砻谷机，中国云南。

巴斯马蒂米是一种产于印度、巴基斯坦和孟加拉国的长粒米，通常是呈白色、富有香气的陈米，因具有爆米花一样的香味和烹饪后蓬松的口感而受到颇高评价。在煮熟之后，巴斯马蒂米仍然能保持粒粒分明。

糯米，可长可短，煮熟后会粘在一起。因此，糯米有时会被当成一种烹饪"工具"，用来"粘"起其他食材。有的人觉得糯米比其他种类的米有更强的饱腹感。长粒糯米存在于泰国、老挝和中国西南部的聚落中。

"本大叔"是一种蒸谷米的品牌。这种米的糠皮中的营养物质会在煮到半熟时被"挤进"谷仁中，因此保留了大概80%的营养价值。"本大叔"是玛氏食品公司一个子公司"每食富"生产制造的全球性米类品牌。

籼稻生长在热带和亚热带地区，占全球水稻贸易总量的 75% 以上。籼米煮干后仍保持粒粒分明。

粳稻通常长在稍微凉快一些的地区，约占全球水稻贸易量的 10%。粳米会有一些黏性，可以很轻松地用筷子夹着吃。

香米，主要是泰国的茉莉香米以及印度、巴基斯坦和孟加拉国的巴斯马蒂米，占全球水稻贸易量的 10% 以上，通常在市场上溢价出售。它们都拥有形状纤长、香气独特、烹饪后口感蓬松的特点。

东南亚产的糯米（glutinous rice）往往被用于制作甜点或者节日菜肴，占世界水稻贸易量的 5%。糯米会被捏成圆柱状和球状来蘸酱和咖喱吃，或者被制成薄饼来卷甜味或咸味的馅料。虽然糯米和小麦、黑麦、大麦中的蛋白质麸质（gluten）的英文拼写很像，但它们其实没有什么联系。

如同许多人想要把这种富有多样性又适应性强的谷类植物进行分类一样，美国农业部（USDA）将其分为四类：籼米、粳米、香米和糯米——不包括爪哇米，因为它介于籼米和粳米之间。美国农业部同时还忽略了一个事实，即在泰国北部、老挝和中国云南，糯米不仅用于制作甜点，也是主食的一部分。在美国，茉莉香米和巴斯马蒂米分别被称为 "Jazzmen" 和 "Texmati"，虽然它们的发音与茉莉香米（Jasmine）和巴斯马蒂米（Basmati）十分相似，但它们都没有被纳入官方的分类系统中，尽管通过游说华盛顿相关部门这种情况可能会改变。

大米中含有两种主要的淀粉：直链淀粉和支链淀粉。两者的含量比例决定了米饭的蓬松度和黏性。糯米的谷粒通常更圆、更短，直链淀粉的含量有限。煮熟后，糯米可以被压制成块状或者团成球状，蘸酱料食用。如果用以面食为主的饮食文化打比方的话，这种饭团和一片法棍面包或者楔形烤馕的作用差不多——都是用来蘸酱汁或搭配肉块、鱼块的。糯米皮也可以卷蔬菜、豆酱或者水果，甚至用来包饺子。

浸泡并蒸熟后的糯米在中国南方、老挝和泰国被当作主食食用。在日本、韩国和中国北方，人们更喜欢不太黏的米饭，然而在印度尼西亚、菲律宾、马来西亚和越南，人们更偏好中等黏性的米饭。这些米属于粳米或爪哇米品种，其口感处于黏糯和蓬松的中间地带。

虽然泰国、越南和美国是世界上排名前三位的籼稻出口国，世界各地的糯米狂热者的数量却与日俱增。糯米饭的另一种形式——甜糯米饭或糯米糍，产生于中国西南地区，且在当地备受喜爱，它在东南亚和日本常被当作布丁或者甜点的外皮。有时候，糯米和非糯米也会混合使用，以获得特殊的风味、质地和一定程度的延展性。当直接食用时，糯米饭有时会被装在烹饪时使用的编织蒸屉里。

短粒米习惯用来制作西班牙海鲜饭，但是中粒米也可以。海鲜饭锅又宽又浅，带有两个把手。先将香料和／或肉类在锅中炒香，再把干的大米加进去翻炒一会儿，最后注入热汤。经典版本通常是放兔肉或者蜗牛；现代的做法则

糯米饭煮熟后放在蒸笼里食用：这笼米饭也是你的餐具。

更多地选用贝类、鸡肉和蔬菜。海鲜饭的传统做法是把不加盖子的锅直接放在室外的火堆或者烤架上烹饪，直到液体全部被收干。没有多余的水分后，锅内的温度开始攀升，这样才能形成锅巴——锅底那一层褐色的酥脆米饭。现代的改良方法则是先把香料和米饭混合腌渍，然后把其他食材和液体倒入锅中，加盖放入烤箱来制作。

　　大米世界的另一端，便是那些被称为巴斯马蒂米或者茉莉香米的长粒米世界。[真正产自印度、巴基斯坦和孟加拉国的巴斯马蒂米（Basmati）和产自泰国的茉莉香米（Jasmine）首字母要大写，这是为了保护产自特定区域的大米。而非这两个特定品种但属相类似品类的长粒米就只能使用小写字母开头的"basmati"和"jasmine"。] 在煮熟之后，这种米可以膨胀到它原来长度的两三倍并且保持粒粒分明的状态。如果要制作白米饭的话，大米会先被浸泡、冲洗干净，然后像煮意大利面一样，在有盖的锅或平底锅中以小火煮或蒸。如果是制作像手抓饭、印度香饭、汤饭和焗饭这些需要在上桌前加入其他食材的餐点，那么就需要用到不同的烹饪技巧了。以手抓饭为例，在与热汤（如高汤）混合之前，大米需要先沁入丰腴的油脂（传统做法是选用肥尾羊的脂肪，波斯的一种珍贵食材）。在烹饪过程的不同阶段还要加入其他的食材。完成整道工序后，你就

街头美食：泰国蕉叶包糯米饭。

能获得一份将鸡肉、葡萄干、鹰嘴豆等食材完美融合的手抓饭。在米饭中间挖个小洞，倒入酥油（澄清的黄油）。接着用面糊或者毛巾把锅密封好，然后用文火慢慢焖出一层锅巴，这种锅底部的褐色脆皮很像西班牙海鲜饭里的锅巴。最后，将分别调味并煮熟的羊肉和饭层层叠叠地摆在盘子上，你就拥有了一份手抓饭。当然，这道菜衍生出了很多变体。在中国，那层酥脆的硬壳叫"锅巴"，韩国叫"nurungji"，在塞内加尔则叫"xoon"。

大米的发展

水稻产业的发展主要得益于 20 世纪 40 年代到 70 年代间展开的绿色革命。农业科学的发展促进了粮食增产，提高了作物的抗病能力，使 10 亿人免于饥饿。1970 年，诺贝尔和平奖首次授予了农业科学家诺曼·博洛格，以表彰他在墨西哥培育出高产、半矮秆的小麦和玉米这一开创性成就。依靠国际水稻研究所的赞助，他的方法后来也被拓展到亚洲的水稻和小麦种植上。这个研究所不但保存了上千个水稻品种的样本，而且一直出资支持提高水稻产量、减少病虫害的相关研究项目。时至今日，国际水稻研究所已经培育出几种高产水稻种子，并在继续研究更高产和低耗水的新品种。在换成高产水稻品种并采用现代化的管理方式后，韩国已经实现了大米的自给自足。而印度也开始使用激光平田技术——用激光来平整土地并划出规整的田垄。

这个办法不仅比人工劳作更加省力，而且灌溉规整的农田也更省水。

灌溉稻田会产生一些沼气——虽然不算主要原因——这种气体也会造成全球变暖。（除了生产供人类消费的粮食这个用途以外）有的土地被用来生产生物燃料和研发转基因（GE）水稻，这些有争议的话题和水稻的未来息息相关。但有的国家虽然在使用现代化的种子，却重新回归了传统的水稻耕种方式（例如，印度尼西亚从20世纪70年代开始就恢复了传统模式）。虽然一些海岸地区（其中很多是水稻种植区）过去10年获得了创纪录的收成，但这些地区已经因为气候变化而被海水淹没了。

保护、保存和种植过去的水稻品种是一种避免未来作物基因由于受到可能的特定环境问题或疾病影响而变得单一的方法。为了避免单一水稻栽培，国际水稻研究所一直保留着一个水稻品种的基因库。

其他关于水稻的科学研究也能解决另外的潜在问题。2010年，位于印度克塔克的中央水稻研究所就培植出了一个含有极低直链淀粉的新品种——"aghonibora"。这种低淀粉大米的优势在于，将它制成速食米饭的耗水量非常低。它只需在温水中浸泡30分钟，就能直接食用了。考虑到气候变暖会使世界的很多地方洪水频发，河漫滩稻也在被研究作为未来可供替代的谷物。因为湄公河三角洲很可能由于海平面上升而消失，因此人们建立了越南湄公河三角洲水稻研究所来培植高产、成熟期短、便于在湄公河三角洲平原以外地区种植的水稻品种。在美国，坐落于阿肯色州斯图加特市（美国大米一半都产自阿肯色州）的水稻研究与推广中心也在努力保持美国大米在国际市场上的竞争力，并且研究适合未来的新品种。

可持续农业的研究和发展将为那些已经以大米为生，以及未来的人们继续提供大米。

［第二章］

旧世界

大约 1,5000 年前，栽培作物的本领让一些靠采集和打猎为生的人形成农耕部落。在众多作物中，像小麦、大麦、小米、高粱和水稻这样的谷物更受欢迎。一些历史学家指出，第一批种植水稻的应该是女人，因为她们的主要工作就是采集，而她们会在水稻生长的河边捕鱼。而另一些历史学家则坚称，水稻最早是一种旱地作物。在自然环境、气候、人类使用和迁徙的共同作用下，水稻的生长区域才慢慢向河岸靠近。

像许多作物一样，水稻被种在森林里的空地上。迁移农业——或者叫刀耕火种的耕种模式——仍然在东南亚或西非的高地上延续着。人们通过控制灌溉来获得更高的产量。他们为了强化产量发明了"耙地"技术，在地下形成了平整坚硬的土层。这可以有效地防止灌溉水很快流失，

水稻穗，每一粒稻谷都清晰可见。

同时瓦解了土壤的内部结构。这让水稻秧苗在杂草丛生的田里获得了立足之地，这个方法可以大大提高有限水源的利用率。人工培植的秧苗会在一至六周后移植到水田里。耙地技术大概率是在印度发明的，后来在中国得以改进并扩展，如今传遍了整个世界。

其他的栽培品种在南亚和次大陆温暖、多雨、潮湿的环境中演化，慢慢适应了不同的地形：水稻是在靠近河流或者河口地带进化而来的；旱稻则长在气候较为干冷的平地或山坡上。在河漫滩上，水稻进化出能够承受水位上升的品种，水稻的穗（带有谷粒的分支花簇）保持在水面上。

以中国为主的故事

大米的历史开始于印度东北部的喜马拉雅山脉的山脚下、东南亚、中国南部和印度尼西亚。"驯化"水稻开始于印度和中国，后来传到韩国、日本、菲律宾、斯里兰卡和印度尼西亚。在中国长江流域南部、泰国的神灵洞[①]、印度北方邦的科尔迪华和韩国的防筑里，考古学家都在出土陶器的碎片上发现了炭化的谷粒。此外，在炭化的米粒附近也发现了吃贮藏米的甲虫化石。距今最早的水稻可以追溯至一万五千多年前。不仅如此，中国人在筑造长城的时

---

① 神灵洞是位于泰国夜丰颂府邦玛帕县的一个山洞。1966 年，美国人类学家、考古学家切斯特·戈尔曼在神灵洞里发掘出了距今约 12,000—7,000 年的和平史前文化遗址。

候，糯米被煮成黏稠的粥状，混以石灰和沙子，用作砂浆涂抹在约重 10 千克的砖块之间。

煮或者蒸可以说是食用米饭最快捷方便的做法。如果能把大米先浸泡一会儿，让它们提前吸收水分保持柔软，在烹饪搅拌的时候，它们就不会轻易碎掉。然而，如果有很多碎的大米，那么你应该做粥。要不然你就干脆把米用杵和臼直接捣成粉（这个步骤也可以用来给稻穗脱壳），然后做薄饼或者米粉来吃。米纸是一种可以食用的"包装纸"，在东南亚和中国西南部比较常见。（可不要把米纸和画家用的宣纸或者其他不可食用的东西搞混了——可食用的米纸是用米粉和水做的，有时也会加入鸡蛋来制成面团；而不可

尼泊尔人割水稻的镰刀上有一个刀锋遮挡物，这样收割稻子的时候，神明就不会知道了。

米粉

食用的宣纸则是用收获过后剩余的秸秆做的。）水稻的茎秆可以用来编篮子、草席和草鞋，或者当燃料。稻壳和米糠油曾被用作动物饲料，而现在被摆放在一些高档的超市里售卖，用来烹饪和拌沙拉。

公元6世纪以来，南方地区就被视为中国的粮仓；而北方则是政治中心，主要负责守卫边疆和巩固集权统治。大米喂饱了士兵，也避免了饥荒。即使在以小麦和小米为主食的北方，稀饭（水煮的米粥或泡饭）从古至今都算得上是一种常见的食物。在世界的任何角落都能找到这种食物，它在韩国被叫作"juk"，在印度被称为"kanji"，在日本被称为"okayu"。一千年后，我们有了意大利烩饭、美国奶油米糊。其实这些都和稀饭很类似，然而，由于它们所使用的大米和烹饪方法，大米在膳食中的作用以及消费者人口特征的不同，这两种汤饭又不同于稀饭。当然，有人会说米布丁根本就是变甜、变稠的稀饭。

在宋朝时期（960—1279），占城稻由今天的越南传入了中国南方。这种熟成快、耐旱的水稻可以达到一年两到三熟，因此很快占据了主导地位。朝廷鼓励稻农（专门种植水稻的农士）在耕种方法或者技术上寻求革新。这些农士受过教育，并会走村串户贯彻落实朝廷颁布的政策。占城稻给农民提供了可靠的粮食保障。在用第一茬水稻给地主交完地租和赋税后，第一茬剩下的和第二茬就都归农民自己了。朝廷还很支持农民去研究能够应对不同气候、海拔和土壤以及增加产量的水稻品种的知识。稻农对他们曾收集、种植、收获、食用、储藏或交易过的特定水稻品种操作非常熟练。杂交水稻也很常见，因为农民会特意保留那些具有理想特征的水稻种子。最终，水稻农业占据了主

腰酸背痛的工作：稻田里的农民们。

导地位，而且种类繁多。当需要更多的土地时，人们就迁移去寻找更适合开发的地区。在蒙古人统治中国后，马可·波罗于公元 1275 年到 1292 年之间来此游历，忽必烈曾在皇宫里以米酒款待他。

在皇帝以及他的朝臣的餐桌上，白米饭绝对算是一种奢侈品。据记载，乾隆皇帝（1736—1795 年在位）统治时期中国烹饪技艺达到一个高峰。朝廷的宴会上通常会有汤、鱼、肉、蔬菜、面条和甜食。白米饭也会作为一些菜肴的配料出现在餐桌上，但它更多是菜肴中的一部分：像塞了糯米的藕或者粽子。宴会后会端上助消化的米粥，当时这是治疗胃不舒服的标准方法（如今也为人们所认可），并且暴饮暴食在宴会上是常态。与季节有关的粥的一个例子是"腊八粥"，这种节庆粥里放入了果脯、坚果和豆子。

在播种季前夕，中国云南的梯田被灌满了水。

在 2010 年，全球生产的水稻提供了世界人口热量摄入总量的 20%，而其中三分之一是中国大米，中国人仅用占其土地总面积 7% 的可耕地养活了世界 25% 的人口。然而，由于世界上生产的大米只有 5%~10% 用于贸易，其中大部分在当地消费，所以任何因气候或政治因素而造成的大米价格波动都会对全球市场上的大米造成不成比例的影响。随着城市化的脚步加快，很多熟练的稻农正在向大城市迁移，尽管机械化对当今的水稻农业产生了一些影响，但熟练稻农的流失令人担忧。

亚洲与贸易

在中国境内外，大米都是通过驳船沿河运输的。大运河长达 1776 千米，将货物从南方的杭州运到北方的北京，也给军队运去了大米。大米也通过驴或骆驼商队沿着各种"丝绸之路"运输。我们最熟悉的"丝绸之路"，从中亚一直延伸到波斯湾和地中海沿岸。南方丝绸之路则东起中国四川省，穿越现在的印度（最早叫巴拉特，后来又改称印度斯坦），西至中亚的巴克特里亚王国。最后，海上丝绸之路以几个港口连接，比如交州（现在的越南北部）和中国广州。海上丝绸之路联结整个中南半岛的海岸，穿过马六甲海峡，直达印度洋和波斯湾。

当然，货船的载重量要比骆驼大得多，特别是运输珍贵的香料、皮草、瓷器、丝织品的时候。其实，每种运输方式都各有利弊，也都有存在的必要。牲畜需要得到喂食和照料，并且可能会生病。驳船非常慢。帆船又要仰仗盛行风（也叫"贸易风"），还有可能因为风暴或者海盗的袭击而沉没。

大米能换来锡（一种制作合金的重要原料，也可以作为其他毒性强的金属的涂层）、杏仁、木材、瓷器、贝壳、象牙、熏香和香料。大米也会被当作货币。即使没有被倒手，一定重量的大米也可以在物物交换时被用作估价标准。可以说大米就是当时的"金本位"。米还会被当作帆船的压舱物，然后在到达港口时被卖掉。因为那时的人们觉得陈米比新米更美味，大米与像小麦这样运输中不宜久存的谷物相比也就更有优势。

人们推测，早在公元13世纪，米粉就被阿拉伯或者印度商人（穆斯林）带到了印度尼西亚和马来半岛。佛教、印度教和伊斯兰教也都影响着米饭的烹饪。米饭是大多数餐食的基本食材，搭配肉、鱼、蔬菜，这些食物过去和现在都是佐以咖喱、叁巴酱和虾酱与罗望子汁调和的辣味调料食用。爪哇产的大米一般用来配咖喱，而甜点会用糯米来制作。

大约在17世纪时，中国就在马来西亚的西岸设立了贸易中心，一些中国人也就迁到了印度尼西亚。马六甲海峡更是在几个世纪以前就建成了仓库和码头等基础设施。这些受保护的水域使中国、印度和阿拉伯海湾的船只得以相遇并进行贸易。到这些海岸旅游的男人经常与他们在那里遇到的马来西亚和印度尼西亚的女人结婚。这些跨国婚姻中的女性被冠以荣誉称号——娘惹，她们将自己的烹饪习惯与中国人的烹饪习惯融合在一起。就这样，中式菜谱、炒锅烹饪和来自马来西亚当地的各式香料融合成大熔炉般的菜肴，马来西亚烹饪继续将马来西亚、中国、印度、娘惹四大传统菜系融合在一起。重辣和发酵虾酱等鲜美的调料抵消了米饭的温和与椰奶的浓郁。由于马来西亚的主要人口是穆斯林，他们不吃猪肉，而中国人吃，印度教徒不吃牛肉，他们都因喜爱吃米饭而联系在一起。以鱼肉香饭为例，从这道类似辣鱼肉盖浇饭的菜式中，你甚至能觅到从阿富汗和北印度传过来的蒙古人做手抓饭的技巧。用洋葱、大蒜、姜、香菜和茴香调过味的鱼片被层层叠叠地码在混合着酥油、豆蔻、丁香和肉桂的长粒米饭上，最后点缀以熟透的番茄丁和厚椰浆，这道菜也是你无法拒绝的美食。其他菜肴则配以原味的蒸饭，辅以香料，如咖喱鱼或醋鸡。就像苏门答腊饮食文化的学者和作家斯里·欧文写的那样，白米饭在以米为主的餐桌上总是最重要的，其他的一切菜肴都要为它让路。

而娘惹风味的另一个代表就是用黑糯米、棕榈糖、香兰叶和椰奶做的黑糯米甜汤（pulut hitam）——配上切好的熟香蕉和浓椰奶，它和大米布丁的相似之处很明显，两个细微的差别分别是糯米本身的黏度使它不再需要加入鸡蛋增稠，以及当地的棕榈糖、香兰叶和椰奶取代了在美国 / 欧洲橱柜里更为常见的精米、白糖和牛奶，无论是新鲜的还是浓缩的。这里的熟香蕉和浓椰奶在美国被换成葡萄干，而法国人用的是果脯和坚果，意大利人更青睐糖渍栗子。

马来西亚有一种很受欢迎的早餐叻沙（lakhsha），这是一种用椰奶调和的辣椒、干虾酱、鸡肉、柠檬香茅和香菜混合的火辣辣的米粉。"lakhsha"这个词来自波斯语，是面条的意思。人们普遍认为，是波斯人在汉朝（前206—公元220）将面条的制作方法传入中国的。

马来西亚版本的米布丁，加入了椰奶和烤腰果。

印度风潮

至少在5000年前，阿富汗和印度北部地区便开始独立培植水稻了。它向西传播至印度河流域，向南传播到印度半岛。公元前2500年左右，恒河附近开始种植水稻，为了躲避来自中亚的蒙古人的侵袭以及寻找适宜耕种的土地，半游牧的猎人和渔民开始了有规律的迁徙。公元前2000年左右，这些印度雅利安人的足迹已经遍及高加索、波斯和兴都库什山脉，最终他们在旁遮普、德里和阿富汗安家落户。尽管印度南边的大米消费量高于北边，但旁遮普的五条河流为印度大部分水稻提供了灌溉水源。受到莫卧儿王朝的影响，一种用奶油、水果和坚果点缀的肉食配饭"pilaus"出现了。如果你往南走，"idlis"和"dosas"——都是发酵类的米食（加酸处理和发酵可以降低细菌污染风险并且延长保质期）——以及木豆（小扁豆）米饭就变得更为常见了。对于克什米尔人来说，只有拿茴香、丁香、肉桂和小豆蔻好好调味过的米饭才能被用来制作手抓饭。而在孟加拉人眼中，"神圣的三位一体"的味道是鱼、米饭和芥末籽油。这样不容置疑的黄金搭配就好像卡真菜系里的芹菜、青葱和青椒，或者法餐里的"mirepoix"（一种基础香料，包括胡萝卜、芹菜和洋葱）。对于喀拉拉邦的居民来说是咖喱叶和椰子香米。而羊肉咖喱配米饭则展现了穆斯林对当地菜式的影响。

当欧洲人在中世纪末开始对亚洲进行殖民统治的时候，大米在印度南部被称为"batty"，在北部叫作"paddy"。这两种称呼都源于梵文里的"Bhakta"一词，意思是煮熟的米饭。一家一户承包小块耕地，却有极高的产量。在丰年，如果谷神满意，对家庭农户显露善意（或者是地主收够了当年的佃租），农民就能有足以糊口的大米，甚至还能把多余的储藏起来或者拿去售卖。

存起来的大米要么是未脱壳的种子，要么是精米，很少是二者中间的状态。因为米糠和胚芽里都富含油脂，在热带很快就会变质。要想得到雪白的大米，需要用杵和臼把糠层和胚去掉。而且米糠还不止一层，所以难免会有些残留在米粒上，除非大米被完全抛光，这样它就变成了精米，可以储存多年。有的人觉得放了几年的陈米味道更好，也更适合烹饪，因为它们经过更彻底的干燥，所以做熟后口感格外蓬松。

人们认为籼稻是从印度次大陆和东南亚迁移至斯里兰卡、马来西亚、印度尼西亚和中国长江以南地区的。爪哇稻在亚洲东南部的高地上生长，而在印度尼西亚则成为低地水稻，漂洋过海来到菲律宾、中国和日本。

印度人还会做膨化的米花来当作零食、早餐或者宗教仪式食品。米花可以直接吃或者就牛奶和白糖（这听起来很像米制的谷物脆，但那个就是真正的膨化食品了）。最常用到膨化米粒的食物就是印度油炸土豆饼（chaat），这是一种极富变化的街头小吃。印度手抓薄脆饭（bhel puri）也是一种在印度和印度移民中很受欢迎的小食，包含膨化大米、土豆、西红柿、薄荷酸辣酱、干脆面、花生、柠檬汁、辣椒和香菜等多种食材。印度出产的大米，大概有一成会被加工成薄片、膨化食品。

伊斯兰教的影响

在公元前 1000 年左右，大米从印度经由阿富汗和波斯传入了中东地区。到了大约公元前 500 年，阿契美尼德人建立的波斯帝国征服了远至印度河的亚洲大陆，同时吞并了希腊、北非、埃及和利比亚。阿拉伯的穆斯林商人通过伊朗和阿富汗将波斯饮食文化带到印度，并且继续前往北非、土耳其、希腊，抵达意大利港口，尤其是威尼斯，随后到达西班牙。

公元 10 世纪，来自非洲西北部的摩尔人大举入侵欧洲南部，并在那里确立了统治，也因此将水稻带上西西里岛、西班牙南部和北非。等到 15 世纪中期，商业化的大米贸易已经在意大利北部开展起来。阿巴斯王国的都城巴格达本来是纯正的伊斯兰美食中心，但随着与西班牙在农业、食材和餐饮上的频繁交流，逐渐演变为伊斯兰世界里一个文化与美食的大熔炉。同样地，西班牙的科尔多瓦也成为伊斯兰文化和美食之都。在这样的混合菜系里，像橄榄、青柠、刺山柑、茄子、玫瑰花瓣、杏、洋蓟、长角豆、藏红花、砂糖、枣、柑橘和胡萝卜等各地食材都被用于烹饪。当地人也会用大麦和鱼做出一种类似鱼露的调味品。长粒米被用于制作各类肉食盖饭（pilaus，这种食物的名字不太固定，也叫"pilaf"、"pilau"或者"pulao"）；短一点或者圆一点的米就被拿来制作这道菜的"远亲"——海鲜饭。大米也会被灌进水果、蔬菜、葡萄叶或者香肠里吃。

公元前 4 世纪初，亚历山大大帝就把水稻从印度带到了希腊。然而，作为一种奢侈品，大米大部分是用来入药的，但偶尔也会出现在宴会上。古希腊的内科医生盖伦和安提姆斯都推荐用山羊奶稀饭治疗胃病，其中的米必须煮得很透。以下是安提姆斯为肠胃不适的患者诊治所用的一则处方：

> 用干净的水煮米。大米煮软后，将水沥干净，再注入羊奶。把锅放于火上，以文火将食物慢慢煮至黏稠固体状。尽快食用，避免放凉，免油免盐。

在 *Ni'matnama-i Nasir al-Din Shah*[①]（约 1500 年）的插图中，侍从们正在给一位波斯苏丹煮米粥。

---

① 一本用波斯语写的中世纪印度食谱。

公元 7 世纪，穆斯林商人把亚洲大米带到地中海。尽管阿拉伯裔的穆斯林已经与中国和其他亚洲小国建立了长达几个世纪的贸易往来，长远的眼光让他们开始把重心移向地中海文明，在农业技术上也逐步向西班牙、西西里岛、意大利、埃及和叙利亚靠近。在埃及，水稻种植在尼罗河沿岸：河流的灌溉和航运优势让埃及成了大米贸易的交通枢纽。罗马、波斯、中国和阿拉伯农业体系中著名的水车工程促进了西班牙巴伦西亚地区、西西里岛和意大利北部波河河谷地区的水稻种植业。筒车这种运水的巨大水轮最早是由罗马人搭建的。固定在每条轮毂上的木桶将水倾倒进大大小小的灌溉水渠。西班牙曾经有过 8000 多个筒车，直到今天，你还能在这个国家找到一些遗留下来的水车残骸。

保存至今的西班牙筒车

［第三章］

新世界

非洲稻（Oryza glaberrima）已经种植了几千年，主要分布在西非沿海的国家、非洲中部和马达加斯加岛。这种红色的水稻最早出现在尼日尔河的三角洲上，属于亚洲稻（O. sativa）以外的另一个重要稻属，而这种非洲的主要作物也与亚洲稻有着明显的差异。这里要特别声明，直到20世纪末，非洲稻才成为学术研究的主要对象。卡尔·林奈①压根儿就没把这种谷物收录进自己的植物学分类系统，而对非洲农业的重要性的故意忽视直接导致了相关历史研究的推迟。直到最近30年，科学家们才对非洲稻燃起兴致，进而出现了许多关于非洲稻的新研究和信息，以及将它与亚洲稻进行的颇有意义的对比研究。

非洲稻对于盐碱物质的耐受性更好，因此更适宜在近海地区用海水灌溉。跟亚洲稻相比，它的谷粒更小，呈深红色，而且有更浓郁的坚果香味。非洲稻的谷穗即使在成熟时也直挺挺地立着，不会像它的亚洲近亲一样长得头重脚轻，因此更便于收割。但是，在用杵臼脱壳时，非洲稻更容易碎；要想获得完好无损的米粒，需要一身熟练而精湛的本领。这种稻米同高粱、小米、山药和秋葵等植物一样，也是非洲的主要农作物。

非洲的大米和当地的奴隶一同被送往英国、葡萄牙、法国和西班牙的殖民地。来自非洲西部"大米海岸"国家和地区（包括冈比亚、安哥拉、几内亚、几内亚比绍、塞

---

① 卡尔·冯·林奈，瑞典生物学家，动植物双名命名法的创立者。

拉利昂和南塞内加尔，以及尼日尔河二角洲）的人因其水稻农业专业知识而受到追捧。这样一个以大米为核心的三角贸易就在大英帝国、西非和西欧之间展开了。

尽管最终大米贸易还是从非洲稻换成了亚洲稻，但非洲稻一在新大陆种植就成了早期殖民地大米产业发展的关键。非洲人对于新大陆"水稻种植"和烹饪的演化进程也有着广泛的影响。无论在耕种、研磨还是烹饪的过程中，非洲稻和亚洲稻都各有优劣。在 20 世纪末，两种稻米的杂交品种非洲新稻（NERICA）被培育出来，也向世人展现了非洲稻对于未来水稻发展的重要意义。较普通水稻而言，这种稻米不仅能在旱地上生长，还更抗虫害，更高产，且其成熟时间更短——从通常的 120~140 天缩短为 90~100 天。

英国殖民地和南卡罗来纳州

已经在巴巴多斯、牙买加和西印度拥有蔗糖种植园的英国商人发现，南卡罗来纳州密布的河流、小溪、湿地、潮汐盆地和亚热带气候非常适合水稻的生长。因为欧洲对高品质的长粒水稻的需求量很大，商人们都明白这种大米是有利可图的。

据说大约在 1685 年，内科医生兼植物学家亨利·伍德沃德博士从一位暂时被困在查尔斯顿的船长那里获得了一小袋大米种子。那正是光秤稻，一种非洲的红色稻米。另个版本的故事则说，在殖民地劳作的妇女和儿童把未脱壳的大米藏在他们的头发里偷运进来。托马斯·杰斐逊甚至把意大利水稻带到了这个地区，他于 1787 年写给政治家爱德华·拉特利奇的信中这样说道：

> 我一定要带出足够多的水稻来给你当种子用，
> 但他们跟我说禁止出口未脱壳的大米；所以我只能
> 尽可能装在大衣和外套口袋里带出去。

英国庄园主知道，来自乔拉、约鲁巴、伊博和曼德族群的非洲人以种植水稻的技术而闻名，从种苗到修建运河和堤坝，他们都很擅长。早在 15 世纪中期，葡萄牙人就认可了他们的这些技能。因此，在奴隶拍卖会上，这些人往往会被叫出最高的价格。从将水稻种子插入湿润的泥塘里以免它们在浇水时被冲走，到锄地、插秧、除草、收割、春米和去壳，奴隶们在地里劳作，形成了一个"田间工厂"，也就是南卡罗来纳州的殖民地水稻种植园。因为他们对大米生产的每个环节都非常熟悉，所以这个种植园的运转就像现代化的工厂一样顺畅，并且有着极高的产量。有的奴隶会自己种点大米、蔬菜和豆类（尽管从 1714 年开始，法律明令禁止劳工个人种植水稻）。有些人会养鸡或者猪、采食、打鱼和捕猎。他们在食品储存室里加入了美国本土的食材，尤其是玉米和辣椒。他们还从主人那里得到口粮，包括每月分配的少量的盐、糖，以及大家最不喜欢吃的猪的部分。

不太理想的是，非洲稻在杵臼中很容易碎掉，导致产量降低。此外，这项工作一般要求男性奴隶承担，因为女性奴隶更擅长碾磨，所以通常会被分派去做家务一类的工作。

因为田里总翻滚着金黄的稻浪，南卡罗来纳州的大米也被称为"卡罗来纳黄金"。它是一种地位非常高的长粒米，尤其是精米，不仅价格昂贵，还在欧洲农业展览会上获得了大奖。在殖民地厨房里使用的英国烹饪书中，也提及过这种稻谷的名字。

美国出版的第一本涉及大米的烹饪书是伊丽莎·史密斯在1742年出版的《完美家庭主妇》。而汉娜·格拉塞在1747年写的《简明烹饪艺术》，是18世纪英国最受欢迎的烹饪图书，在殖民家庭中非常有影响力。其中包括布丁、手抓饭、汤、咖喱和煎饼的食谱，它们都是用大米做的。在著名的社交名流莎拉·拉特利奇于1847年著的《卡罗来纳的家庭主妇》里，出现了一道米饭食谱——烩菜豆饭。通过在书里加入这样的食谱，拉特利奇将奴隶烹饪顺利发展成"南方烹饪"。不过，大米仍然是个复杂的话题，不容易讨论，就像比顿太太在1861年出版的《家庭管理之书》中写到的那样：

> 大米的种类：在进入我们市场的大米中，孟加拉产的糙米有着粗糙的棕红色外皮，却粒大味甜；它不容易从外壳中分离出来，但当地人对这种味道

情有独钟。巴特那大米在欧洲更受赞誉，其品质也是非常好的；它是小粒米，相当长而结实，颜色洁白如雪。卡罗来纳的大米被认为是顶级的，同时在伦敦也是最昂贵的。

下面这则出自 J. M. 桑德森 1846 年版的《烹饪完全指南》的菜谱，反映了英法菜系对于米饭在烹饪中使用的早期影响，这些知识很可能是从中世纪制作"法式奶冻"的经验中转化而来的。这可能源于阿拉伯人把杏仁捣碎成"奶"的方法，再加入米或者米粉和糖，配上鸡丝和玫瑰水。在具体的烹饪和命名中，这种布丁变化颇多。

米布丁：取一杯卡罗来纳大米和七杯牛奶；把锅放水里隔水煮，保持水沸腾的状态直至锅里的液体变稠；然后加糖，最后再撒上一盎司①的甜杏仁碎。

一些女奴被安排在"公共大厨房"工作。她们大声朗读英国食谱，希望她们的厨师能够通过这种方式记住所需食材和做法，以准备食物。尽管这些食谱起源于欧洲，但非洲的原材料和烹饪技巧（例如油炸）影响了最后的准备工作。山药、茄子、秋葵、黑眼豆、小米、绿叶菜、西瓜、南瓜、芝麻、红薯、可乐果和高粱都源于非洲。

---

① 重量单位，1 盎司等于 28.3495 克。

RICE CAKE.

1772. INGREDIENTS.—$\frac{1}{2}$ lb. of ground rice, $\frac{1}{2}$ lb. of flour, $\frac{1}{2}$ lb. of loaf sugar, 9 eggs, 20 drops of essence of lemon, or the rind of 1 lemon, $\frac{1}{4}$ lb. of butter.

Mode.—Separate the whites from the yolks of the eggs; whisk them both well, and add to the latter the butter beaten to a cream. Stir in the flour, rice, and lemon (if the rind is used, it must be very finely minced), and beat the mixture well; then add the whites of the eggs, beat the cake again for some time, put it into a buttered mould or tin, and bake it for nearly 1½ hour. It may be flavoured with essence of almonds, when this is preferred.

CAKE-MOULD.

Time.—Nearly 1½ hour. *Average cost,* 1s. 6d.

Seasonable at any time.

比顿太太的米糕食谱

猪的一些精选部位是专供种植园主人及其家人享用的，而奴隶们使用猪蹄、猪头、猪下水、肋排、咸猪肉、熏猪肉和大肠当作调味品或配菜，大多是用来搭配主食米饭。今天，这些食材和食谱仍然是美国南部食物中不可或缺的一部分，不仅是黑人家庭，这些菜甚至会出现在一些白人的餐桌上。奴隶们偶尔会瞒着种植园主偷偷做蜂蜜米糕，这是他们穆斯林的传统。塞内加尔和尼日利亚的奴隶通常是穆斯林，因为禁食猪肉，他们会用风干的牛肉制作他们自己的"烩菜豆饭"。

来自沿海低地、南卡罗来纳州和佐治亚州海岛奴隶的后代，被称为古拉族（Gullahs、Geechees），他们以大米作为三餐的主食。"古拉"的本意，就是指晚餐吃米饭的人。白米饭有时会配着牡蛎、虾、鱼、猪肉和禽类来吃。而更为常见的做法是米饭配绿叶菜和秋葵，有点像塞拉利昂的菜叶炖饭（plasas）或者大米秋葵汤。红米饭可以配上用秋葵、鱼、西红柿和辣椒炖的秋葵浓汤（这道菜的名字取了班图语"nkombo"，意为秋葵），类似于西非辣椒炖肉饭。它被誉为"南卡罗来纳的招牌菜"，也是典型的古拉饮食。

到1690年，水稻已经成为殖民地主要的经济作物。这个产业的发展速度是卡罗来纳州水稻文化的显著特征。在1700年的时候，大概有60万磅的大米从查尔斯顿港运往英

TO BE SOLD on board the Ship *Bance-Island*, on tuesday the 6th of *May* next, at *Ashley-Ferry*; a choice cargo of about 250 fine healthy

NEGROES,

just arrived from the Windward & Rice Coast. —The utmost care has already been taken, and shall be continued, to keep them free from the least danger of being infected with the SMALL-POX, no boat having been on board, and all other communication with people from *Charles-Town* prevented.

Austin, Laurens, & Appleby.

N. B. Full one Half of the above Negroes have had the SMALL-POX in their own Country.

查尔斯顿奴隶拍卖的一则简介，约18世纪80年代。

国和西印度群岛。到1710年就出口了150万磅；而1740年时，出口了4300万磅。这些数量众多的装米货船也因此成了海盗的目标，经常在横渡大西洋的时候被洗劫、被击沉。直到英国派出海军舰艇保护，大米出口才变得安全且有规模一些。1771年时，大约有6000万磅大米通过征收关税的英国向欧洲其他地方输送。1789年，出口贸易达到了8000万磅，为英国带来了源源不断的财富。下面这个表格就展现了大米价格从1772年到1809年的攀升情况。

美国的大米价格（1772—1809）

时期	米价（美分／磅）
1722—1729	1.40
1730—1739	1.64
1740—1749	1.18
1750—1759	1.56
1760—1769	1.58
1770—1779	1.87
1780—1789	3.15
1790—1799	2.73
1800—1809	3.81

在1750年，南卡罗来纳州的一名种植园主麦克奥·约翰斯通发明了依靠潮汐灌溉水稻田的方法，大大增加了可

耕种土地的面积。1767年，乔纳森·卢卡斯又发明了水力磨坊，减少了人力需求，而且增加了全麦谷物的出产。

在美国南北战争期间（1861—1865），种植园被大量摧毁。大米产业开始向西迁移到阿肯色州、路易斯安那州和得克萨斯州大草原，因为其有平坦的土地和机器（并非人力），降低了生产成本。与此同时，猪大肠、羽衣甘蓝、猪蹄髈、玉米粉、黑眼豆和米饭随着被解放或逃亡的黑奴来到北方的城市。底特律、芝加哥、纽约和其他城市的非裔美国人人口开始增长。1964年，"灵魂食物"这个词第一次

在开普菲尔河种植园里劳作的非裔美国人，北卡罗来纳州，1866年，木版画。南北战争之后，大多数奴隶可以自由离开，也有一些奴隶作为受雇雇员留了下来。

出现在报纸上，用以形容这些饱含民族自信与民族情感的食材和烹饪方法。

尽管水稻种植在南卡罗来纳州又存续了80年，但产量很少，并在1920年左右结束。2000年，安森·米尔斯收获了第一批祖传水稻，现在他正在种植其他祖传水稻。南卡罗来纳州查尔斯顿的卡罗来纳金水稻基金会也于2011年2月宣布长粒香米的到来。时至今日，卡罗来纳金牌水稻和查尔斯顿金牌水稻都还在生产。

路易斯安那州、大草原和加利福尼亚州

在法国人与西班牙人为了路易斯安那州的领土争得不可开交的时候，大米烹饪逐渐发展起来。"克里奥尔"和"卡真"是指那些融合了欧洲、加勒比和非洲文化背景的人、原材料和饮食体系。克里奥尔人一般是指那些在西班牙旧殖民地出生的、有西班牙裔双亲的后代。后来，在路易斯安那州出生的法国移民的小孩和奴隶后代也被归到了这个定义里。卡真人则是指那些在1755年英国接管加拿大地区后，被驱逐出阿卡迪亚的法国殖民者。这些阿卡迪亚人（或者叫"卡真人"）一路南迁，大部分在新奥尔良的河口一带登陆。克里奥尔烹饪通常与较富裕的新奥尔良联系在一起，卡真风味则稍晚一些才为人所知，因为那里的河口很难到达，那里的居民更贫穷，菜肴普遍偏辣。

在卡真人和克里奥尔人的食谱中，米饭可以和秋葵浓

汤组合出不同的花样。每个食谱都可能起源于克里奥尔或卡真，这取决于你与谁交谈。而关于这道菜配方的争论就更为火爆了。在克里奥尔的版本里，你很可能会吃到虾、香肠和鸡肉，而白米饭不仅能缓解辣味，还能增加这顿饭的经济价值。如果是卡真风格的浓汤，小龙虾、螃蟹，甚至松鼠、猪肉肠和辣椒都能用来配饭吃。秋葵浓汤是一种将所有的味道都"结合"在一起的菜肴，这点很像手抓饭和海鲜饭，只是更浓稠。

另一道标志性的路易斯安那州美食当属"红豆米饭"，它是烩菜豆饭很多种变体中的一种。这道菜往往蕴含着丰富的地域特色：牙买加人会放胭脂树橙色素，而在多米尼加共和国人们则加入椰汁。通常情况下，烩菜豆饭是庆祝新年的传统食品，而红豆米饭则是洗衣日的固定菜肴。豆子

小龙虾秋葵浓汤：可别忘记左下角露出的米饭！

会和猪蹄髈一起焖熟，然后和白米饭一起吃。传统的吃法还会配上点青菜，然后把"菜汤"也一并喝下（就是做绿叶菜后锅中剩下的美味汤汁）。

当意大利人、德国人和南斯拉夫人移民到路易斯安那州后，他们也把自己的风格加入到克里奥尔/卡真人的文化大熔炉中。在西班牙探险家赫尔南多·德·索托（约1496—1542）称霸的年代，就有意大利人从意大利北部搬到路易斯安那州；德国人则是在1718年新奥尔良建立的时候移民过来的。很多德国人都陆续迁到了北面更平坦的土地做农民，先是种小麦，后来是大米。从当地的沙拉就可以看出德国文化不小的影响力，这道菜兼具德式热土豆沙拉的调味和新世界新奇的改良：培根、糖、苹果醋、芹菜籽

米饭和豆类：豪华版的烩菜豆饭。

和切碎的甜椒、青椒和洋葱被一股脑儿地混入热腾腾的米饭里，再盖上一瓣切好的水煮蛋。白香肠则体现了传统与新食材的结合：这种用碎猪肉、大米和香料灌制的香肠既保留了德国香肠制作的传统工艺，又依据路易斯安那州的农产品、勤俭节约的美德和口味偏好做出了调整。

在19世纪末期，一位农民在阿肯色州种了三英亩的水稻。那里土地很平整，也能承受重新装配的重型机器（原先是用来种小麦的）。磨坊和灌溉技术也有了进步。阿肯色州成为美国的"米篮子"，正如在州歌中唱的那样："田地里盈车嘉穗。"当地人用大篷货车把稻谷送往得克萨斯州和密苏里州，而1881年通车的路易斯安那西铁路则把得克萨斯州的奥兰治和路易斯安那州的拉斐特串联起来，让中西部的麦农和路易斯安那、阿肯色州和得克萨斯州草原上的米农得以沟通。仅用了30年（1880—1910），这片政府评估员断言毫无农业价值的荒地就一跃成为全国最富饶高产的稻田。

然而，中西部的新移民需要去学习如何做出松软的米饭。南部大米种植者协会印刷了带有菜谱的小册子。在1912年，一个叫《克里奥尔妈咪米饭食谱》的小本子里有秋葵浓汤、什锦饭和米豆腐的菜谱。对于来自德国、捷克和斯堪的纳维亚的"土豆信徒"，大米的营养价值也在宣传之列——"大米和豆类"比"小麦和肉类"更适合你！

美国现代的大米产业主要集中在阿肯色州、加利福尼亚州、路易斯安那州、密西西比州、密苏里州和得克萨斯州。

这些地区大多种植长粒水稻，阿肯色州在产量上遥遥领先。就在同一片土地上，产量在过去30年间增长了60%。仅2010年一年，美国就生产了8300万千克长粒米、2600万千克中粒米和不到150万千克的短粒米。

　　萨克拉门托附近的农民发现，水稻比其他作物的长势更好。1920年以来，加州开始了规模化、商品化水稻产业的拓展；现代化的"水稻栽培"方法主要包括在激光平整过的田垄间使用飞机播种、喷洒杀虫剂和化肥等新方法。大约90%的加州大米是中粒稻，它们通常被称为"加州玫瑰"。这个品种的大米杂质少、味道温和、略有黏性，几乎可以做寿司、海鲜饭、亚洲菜等任何米类料理。除了"加州玫瑰"，这里种植和进口的大米也包括有机的短粒米和长粒米，或者像"禁米"黑米、竹香米和不丹红米这样古老的品种。近来，有机稻农场逐步成为加州大米产业中的重

纳蒂亚·盖尔西亚·波拉斯于2010年在哈瓦那绘制的当代艺术品，画中是一位端着米饭的古巴女人。

要一环。

在 1919 年，大约有 100 万英亩的土地被用来栽种水稻，其产量大致为每英亩 1100 磅。到了 2010 年，种植面积已经拓展到 300 万英亩，最高产量为每英亩 6500 磅。1970 年，水稻作物的经济价值为 5 亿美元，2010 年为 30 亿美元。本地销售和出口各占大米总产量的一半。

西班牙、秘鲁和古巴

在 1849—1874 年之间，大约有 10 万名被称为"苦力"的中国雇佣工人在签订了为期八年的合约后来到位于古巴和秘鲁的西班牙殖民地。他们大多是在种植园或沿海农场里工作，或者做家仆。这主要是因为，在 19 世纪 50 年代中期，秘鲁和古巴纷纷独立，奴隶制是非法的，中国苦力已经取代奴隶成为契约劳工。他们要求把大米作为薪酬的一部分。刚开始亚洲稻需要依赖进口，后来便沿着沿海水道得以种植。到了 19 世纪 70 年代，逃脱并获得自由的苦力移居到秘鲁亚马孙地区，也把水稻、豆类和其他作物带到了这片土地。20 世纪中期，在秘鲁的华裔纷纷聚集在利马的中央市场周围，那里也被称为利马唐人街（el barrio chino）。如今，中餐馆（chifas）成了中国与秘鲁文化交融的标志。

1857 年抵达古巴的第一批中国人和非洲奴隶、当地土著一起劳作。中非通婚的现象时有发生（与西班牙人结婚是被禁止的）。最终一共有 12.5 万名苦力前往古巴。非洲和中国的传统饭食在烹饪中相遇，发展出一些融合菜；与此同时，中式古巴菜也在拉丁美洲最大的华人群体之一——利马唐人街——中出现。在哈瓦那，既有人吃着中国南方的传统早餐稀饭，又有人在享用有着拉丁风格的黑豆米饭（被戏称为"摩尔人和基督徒"）。在1959年古巴革命发生后，大多数中国人迁居到迈阿密和纽约，但古巴式的中餐依然在这个国家蓬勃发展。

墨西哥

16世纪20年代，西班牙征服者把水稻引入墨西哥韦拉克鲁斯。毗邻墨西哥湾的地理位置使这片区域成为理想的水稻种植地。在坎佩切湾，甚至可以实现水稻一季两收。大米被引入日常饮食中，用西红柿代替藏红花，常常是先

西班牙鸡肉饭

用油炒一下米饭，再加入高汤和其他原料。

　　首先用油炒米饭被证实有利于烹饪后米粒保持松散，因为是中粒稻，往往比较黏。这种先用油炒再加汤焖的阿拉伯式／西班牙式做法随着殖民势力的发展传到殖民地。这里同时种植了长粒稻和短粒稻。亚洲稻横渡太平洋来到此地，非洲稻也经由西班牙引入。最终，前者在这片土地上称霸，成为当地更经常种植的品种。

　　大米和豆子这对经典拍档并没有很快在这个国家的主流餐饮中兴盛起来。瑞克·贝莱斯（墨西哥烹饪厨师和研究饮食文化的历史学家）指出，大米一旦被纳入传统的膳食中，就会被更全面地接纳。通常用中粒米或者长粒米制作

欧洽塔：炎热天气里必备的清凉饮品。

的墨西哥砂锅（sopa seca），是指一种汤汁全部被收干的饭食，如同韦拉克鲁斯的大米和海鲜制品。湿汤（Sopas aguadas），如同字面意思，是一种炖菜或者浓汤。对于那些无甜不欢的人来说,墨西哥版的米布丁——牛奶米饭（arroz con leche）——是个不错的选择。还有一种清凉的消暑佳饮欧洽塔，是将碾碎的米浸泡过滤后，再用杏仁或肉桂调味而成。

葡萄牙和巴西

1530 年，满载奴隶的荷兰船队穿过佛得角群岛，驶向巴西大陆，也把水稻带上了巴西东海岸的巴伊亚。从苏里南到卡宴，人们都知道女奴和小孩偷偷在头发中藏水稻种子的故事。非洲稻原本是奴隶船开到巴西一路上的粮食供给。然而到了 1550 年，稻米已经是里约热内卢公开贩售的商品了。1618 年时，大米成了巴西甘蔗种植园里奴隶的主要食物。

为了得到巴西的控制权，葡萄牙和荷兰的军队在 17 世纪一直不断交火。随着种植园的发展，越来越多的奴隶从非洲被运了过来。亚洲稻在 1766 年的时候被移栽到这里，成为出口农作物。到了 1781 年，葡萄牙人消耗的大米已经全部产自巴西了。巴西在 1822 年独立以后继续发展水稻种植业，米在巴西人的日常饮食中占据中心地位。随着城市的发展，大米的需求量也在增加。这也部分解释了为什么用豆子、大米、烟熏或者烤制的肉类、羽衣甘蓝、橙子片和烘烤过的木薯粉制作的腓秀雅杂黑豆炖饭全餐（feijoada completa）会被当作国民美食。虽然近年来吃木薯的人变少了，但这种食物原本也是巴西人的主食之一。

成千上万的逃亡奴隶在茂密的马拉尼昂雨林里建立起"基伦博"这个落脚处。他们在那里种植水稻、玉米、木薯和香蕉，也会捕鱼和狩猎陆地上其他动物。在巴西于1888年废止奴隶制之后，一些人搬去了城市，而剩下的留在了"基伦博"，他们的后代直到今天还在同样的土地上种植水稻。

当19世纪末奴隶制在拉丁美洲合法终结的时候，殖民产业、大农庄、种植园和铁路建设业都急需补充新鲜劳动力。满载亚洲移民的船队就这样有序地驶向拉美和南美，以中国、日本和东印度的男性劳工为主。日本人集中在秘鲁和巴西落脚，印度人投奔了英国的西印度公司。大量的中国男劳工则散落在巴西、秘鲁、古巴和墨西哥，并与当地的妇女结婚，逐渐习惯了当地的饮食习惯。

在15世纪末，葡萄牙人把"几内亚水稻"（非洲稻）进口到了欧洲；它还不是主食，需求量却很大。大约在18世纪30年代，因为米饭是在圣日用来搭配鱼肉的重要食物（每年大概有一百多天都是圣日），葡萄牙人不得不从意大利和南卡罗来纳进口水稻，来满足欧洲天主教族群与日俱增的稻米需求。欧洲持续性的粮食短缺造成了水稻进口量的攀升，这使得在巴西种植水稻成了一个貌似可行的解决方案。正是基于这个构想，亚洲水稻被引入南美，成为巴

哥斯达黎加的米粉蒸肉

西的出口商品。非洲水稻最早的发源地其实是亚洲（连林奈也同意这个说法）。在非洲，大米是由葡萄牙商人带到几内亚海岸的（因此得名"几内亚大米"）。20世纪的植物学研究证实，非洲稻的种植至少在4500年前就已经有迹可循了，远早于欧洲人在这片"黑暗大陆"上的冒险。法属西非地区的红色水稻其实是早就生长在那里的独立品种，而不是像欧洲中心主义认为的那样，是欧洲人将水稻文化带到了西非和美洲。在1766年，当"卡罗来纳白水稻"抵达葡萄牙并被送往巴西栽培时，源自非洲的意大利稻其实已经在巴西生根发芽了。虽然砻谷机去除了稻米最外层的谷壳，但因为机械碾磨机的短缺，最传统的杵和臼直到1774年还在使用。鸟类也是个大麻烦，它们会使得一个品种的谷粒掉到另一个品种的田里，让谷类分离变得异常艰难。最终，非洲稻被淘汰掉，亚洲稻成为主流。

英国和印度

英国和印度的贸易最早开始于 17 世纪，在英属印度时期（1858—1947）达到了顶峰。两国之间的美食交流则主要是导向英国一方的，印度移民也大大促进了这一进程。米最早是由从印度回来的盎格鲁官员和留在英国的印度海员带入英国的。慢慢出现的印度餐馆、外卖以及非常方便的印度食物半成品让英式印度菜得到了进一步普及。

19 世纪中期以后，英国成了大米运往欧洲港口之前征税的门户。在南北战争爆发后，美国切断了先前英国殖民地的大米出口贸易，亚洲稻变得越发重要。欧洲的革命也提高了对于水稻的需求。一些没有出口的水稻会用来供给印度移民，以及之后陆续涌入的中国、东南亚和非裔加勒比移民。到了 20 世纪初，在印度的军队统帅和行政人员已经自称盎格鲁-印度人了。"Nabobs"就是对富有的英国人的称呼，他们在英国统治时期任职，1947 年印度独立后回国。他们通常会把自己的厨师也带回来。印度的海员，特别是船上的孟加拉裔厨师，在 20 世纪初来到英国后往往会开个小饭馆。从咖喱馆子、小酒吧到维拉萨米（这是一家由盎格鲁-印度人爱德华·帕尔默在 1926 年开设的印度餐厅，目前仍然经营良好），英国口味的印度菜很常见。诚然，

盎格鲁-印度式早饭：维多利亚时期的版画描绘了印度人及盎格鲁-印度人的用餐礼节和习惯——有饭有鱼。

两种菜式不太一样，米饭则是连接两者最好的桥梁，也正是如此，"咖喱饭"成为最受欢迎的一道菜。"咖喱饭"里配的大多是未经调味过的白米饭。芬芳四溢的手抓饭和印度香饭、豆蔻味的米布丁和配烤串的藏红花米饭，这些菜肴也是盎格鲁-印度文化的一部分。

印度餐馆最开始的目标客人是移民过来的同胞，但很

手抓饭

印度的膨化大米混蜜糖，这难道不是米糖的起源吗？

快它们便吸引了那些舍不得英属印度的男男女女。随着人们对印度风味的渴望越来越强烈，复刻印度美食成为常态。咖喱鸡块就是英式发明中最著名的一道菜。

在现在的超市和百货商场里，罐装的酸辣酱、腌菜、调和咖喱、玛萨拉酱之类的现成酱汁和咖喱肉几乎随处可见。无论是巴特那米还是孟加拉米，抑或是巴斯马蒂米，也都唾手可得。在冷柜里也有成堆的自热食品和那些可以连同袋子一起煮熟或者微波加热后可以直接扣在米饭上吃的方便食品。

然而，这些英国风味的印度食品并不能充分解释为什么这种异域美食会大受欢迎。事实上，早在中世纪，英国（至少在上流社会中）就有使用香料的传统，这或许有助于解释为什么英国人能够接受如此浓烈的调味料，尽管这个观点存在争议。在第二次世界大战后，英国的印度餐馆数量开始飙升，这很有可能是受到持续至20世纪50年代的战时配额制的影响。这样的小餐厅成了移民在融入社会时维持生计的途径。有意思的是，柯林·泰勒·森在《咖喱全球史》（2009）中写到，"咖喱"一词应该只用来形容那些在英式印度厨房里诞生的食物。那米饭又该置身何地呢？虽然它极少被提及，却是餐桌上的常客。这样一粒粒米垒成的白色山丘，是促进盎格鲁–印度美食在英国发展的重要推手。

鱼蛋烩饭（英语拼写为"kedgeree"）就是一道出色的融合菜。这种在印度语里叫作"khichri"的扁豆烩饭是数百万印度人的日常食物。然而，英国人觉得这样单调的吃食未免有点拿不上台面，所以就加入了熏鱼和水煮蛋来提升它的地位，这是中产阶级对动物性蛋白渴望的早期迹象。直到今天，它仍是优雅英式午餐的菜品之一。

印度尼西亚

从15世纪中期开始，荷兰和葡萄牙人就因为对马鲁古群岛（最早叫香料群岛，也被称为摩鹿加群岛）、印度尼西亚其他岛屿和斯里兰卡的控制权一直纷争不断。马鲁古群岛出产的胡椒、肉豆蔻籽、丁香、肉豆蔻皮和姜黄都是利润极高的货物。荷兰在1602年建立了东印度公司以更好地控制贸易，也因此产生了主导影响力。和英国人在印度的情形相似，殖民者与被殖民者划清界限的一个重要表现就是二者在餐桌上的分别。以荷兰控制的爪哇为例，岛上土著的传统食物是大米、蔬菜和汤。基于此，荷兰人为了让自己的用餐体验更加欧洲化（或者说更加复杂），发明了奢侈的"印尼饭宴"（rijsttafel，其实是饭桌的意思）——在米饭旁摆上许多小碟子装的生食、熟食、酱汁、调味料，甚至是炸香蕉。印尼饭宴是一种炫耀阶级优势的菜肴，成为殖民者传统的周日美食，但它和印度尼西亚饮食习惯的联系可以说是微乎其微。对于荷兰人而言，令人眼花缭乱的配菜正是关键所在：丰富的菜品彰显了他们殖民者的高贵血脉，而米饭只是个好吃的佐餐或者清口佳品。直到今天，印尼饭宴专门店在荷兰还是很流行，吸引着大批本国和印尼裔的食客。但在印度尼西亚，却很少能看到当地人做印尼饭宴来吃。

消费者的兴起

米为一，烹之以众法。

——斯瓦希里谚语

到 2050 年，城市人口将会突破 90 亿大关，约占世界总人口的 70%。城市化改变了人们获取和食用大米的方式，而水稻的供应和消费模式也已有所变化，以适应这些趋势。无论是在家还是上班，又或者在美食广场、公司食堂或餐厅里，方便快捷才是关键。尽管越来越多的中产阶级更青睐补充动物性蛋白质，同时减少淀粉类主食的摄入，但是，平价的大米依然是超市货架上不可或缺的产品。

从大型仓储式超市到街角的小杂货铺，罐装、袋装、盒装、冷藏、冷冻和需要微波加热的米制速食品随处可见。从精米、黑米到菰米（它不算真正的稻米，因为它不是稻属而是一种菰属的半年生水生植物），一部分米类半

My man likes something unexpected now and then. That's why I serve him rice.

Rice is full of tasty surprises. It's as quick and easy to vary as adding chopped chives. Or topping with paprika. Or tossing with crumbled bacon and sour cream. Or shredded Cheddar cheese. Or toasted almonds. In fact, there are so many ways to vary rice, you could have it every day for a year and not repeat yourself. That's pretty unexpected from a little carton or package of rice.

Va·rice·ity

For free booklet "Rice Ideas Man Like," write Rice Council of America, Box 22802, Houston, Texas 77027.

大约 20 世纪 70 年代大米委员会的广告上写着："我的男人喜欢惊喜……"果然，打打性感牌，大米也好卖！

成品成了城市消费者准备正餐的必需品。这一章主要聚焦美国的大米消费，但需要注意的是，以下提及的每种产品都有着更古老、更遥远的渊源，它们都发源自更田园的制作方式，并且在世界的其他地方也能找到类似产品。

移民的影响力

美国在 19 世纪后半叶进行了领土扩张，随着得克萨斯州、加利福尼亚州和西南部地区被吞并、攻占或购买，生活在那里的墨西哥人被默认归为"美国人"。就在这个阶段，用洋葱、青椒和西红柿做成的"西班牙饭"（除这些之外，所添加的其他食材都更有争议）普及开来。自西班牙人于 16 世纪 20 年代踏上墨西哥的土地后，米饭就纳入了当地的日常饮食（西班牙人同时还会经太平洋运过来菲律宾的大米）。因此，今天你可以轻易买到盒装的"西班牙饭"。

到了 20 世纪初，美国的意大利移民群体已颇有建树。本是意大利家庭美食的烩饭逐渐出现在餐厅里，现在甚至能买到速食的。美国的阿尔博里奥米[①]主要生长在加利福尼亚州、密苏里州和阿肯色州。我最近还买到了一盒意大利奶酪烩饭，大概是受到了奶酪意面的启发。它的主要原料是加利福尼亚州有机的阿尔博里奥米，做法也简单快捷，只用到黄油和牛奶。米、帕玛森奶酪、盐、奶粉、香草、

---

① 阿尔博里奥米：意大利圆米，产自意大利波河平原的阿尔博里奥小镇，并以此命名。

香料和油这些其他的材料，全都处埋好放在盒子里了。烹饪时间是 20 分钟，不超过做传统烩饭一半的时间。四人份只需花费 3.49 美元。

美国的亚裔人口将在 2020 年达到 2000 万。近期的亚洲移民主要来自中国的香港、台北、福建，以及马来西亚、菲律宾和南亚地区。虽然这些移民的后代很快就习惯了美式饮食，而且越来越常吃快餐，但他们仍然经常在家里做传统的米饭，尤其在周末聚会和节假日时。当传统被同化，混搭观念就产生了。在超市的货架上，亚洲饮食进军速食业，象征着另一种演变。柠檬草米粉和糯米寿司卷都是今天常见的方便食品。这些产品在美国几乎随处可见，特别是在加利福尼亚州、纽约州和得克萨斯州这些亚洲移民比例很高的州。

运输和贸易

到了 19 世纪 50 年代，因为火车和卡车的发展，货运变得更加快捷且便宜。冷藏运输技术使半成品食物得以完成从产地到卖家的点对点运输。随着带有冷藏舱的飞机和货船也加入了混合运输系统，这一通过控温来确保食材新鲜和卫生的"冷链"将米制品从工厂带到了各地的冷冻展柜，也让这些经过处理的食物有了更长的保质期，冷冻食品的品质得到了大幅提升。那些只需加热就能立即食用的餐食成了超市食品区不可或缺的一部分。

世界上一半的大米都被运往城市。工厂和加工车间大都位于大米产地和消费市场之间，有些现代大米生产工厂甚至就建在之前的稻田上。大米已经由一种用来果腹的主食发展成迎合消费者对于口味、健康、价值和便捷的追求的加工食品。另外，40% 的美国大米都被用来酿造啤酒。制作清酒、味噌和日式白米醋也都需要用到发酵的大米。

最近，我在纽约家附近的超市里买到了一袋蒂尔达牌泰国茉莉香米。这种米是在泰国生产、英国脱壳包装，最后在美国和其他地区出售的。尽管补助条款和国际贸易协议总是引来争议，但当代全球化的贸易环境还是鼓励这种通过出口经济创造现金流的跨国生产分配模式。在米价低的时候，一些卖家可能会囤积居奇。泰国是一个重要的大米出口国，尤其盛产高档的茉莉香米。就像英国在殖民时期做的那样，现在的泰国和美国政府都会对进口的大米加征关税。

上班族的饮食习惯也发生了变化。他们不再像往常一样规律地回家吃午饭，而是带饭上班或者直接买工作餐吃。无论是铝制饭盒、环保纸盒、翻盖塑料盒还是中餐外卖盒，都能使工作餐中的食物保质保温。寿司和冷食米沙拉也随处可见，现在甚至还会放在一次性便当盒中贩售了。如今很多办公室都配备微波炉，可以简单制作或者加热从家里带来或从清真餐馆、印度街边摊、餐车、超市熟食区、公司食堂或餐厅买来的午餐。外卖也是上班族的标配之一。

无论是在洛杉矶、多伦多还是伦敦，"蒂芬快餐"①都会被送往印度程序员的办公桌。和印度的城市一样，每天都有人点这种装有米饭、豆类和咖喱的分层式盎格鲁-印度午餐便当。虽然餐盒本身不需要回收，但点餐的人可以把前一天的餐盒交与外卖员送回，表示需要预订次日的午餐。或许，这会成为一种新风潮？

创　新

在19世纪40年代，拿破仑举办了一场发明大赛，旨在寻找可运至战区的更安全、保质时间更长的物资，而冠军得主正是凭其设计的罐头食品参赛的法国化学家阿佩尔。在第一次世界大战期间，美军的主要食品之一正是罐装的熟米饭，他们喜欢加入汤和碎牛肉来食用。

然而在19世纪末，制造罐头用的镀锡铁是军用急需物资，无法用在食品制造领域。因此，开发新的军粮储存办法成了迫在眉睫的任务。1879年，罗伯特·盖尔发明了可折叠的纸饭盒。纸盒还有突出的优势：它不仅自身密闭性好、非常轻便，而且里面放的食物可以更长时间地保持干燥和新鲜。包装设计和广告宣传也为它吸引来大批买家。

冷藏技术的革新给消费者带来了更广泛的选择。无菌包装和真空密封包装袋让速食米饭的口感和味道更上一层

---

① 蒂芬快餐：印度连锁外卖品牌。"蒂芬"为印度语中的英语外来词，可释义为午餐。

楼。巴氏消毒法和其他食品安全技术为大众化的食品制造开辟了新的道路，受益的群体不仅涉及在超市里采购的平民百姓，还包括以配有加热材料的自热食品为生的军队。学校、医院和监狱都受益于这些袋装、罐装、冷冻或可微波加热的速食米制品。这个变化甚至直接反映在了美军手册上。1906年的美军手册记录了罐头米饭，而到了2006年，军人已经可以自由选择是吃即食墨西哥饭、炒饭还是圣达菲豆饭了。

尽管本土主义还是存在，但我们早已不是地里长什么就吃什么的古代人，我们考虑便捷和利益，吃着从另一个国家生产和出口的食物。我们从网上下单购买大米或米制品，它们可以被运往世界上的任何一个地方。我们往往把这种现象和发达国家绑定在一起，但事实上，这样的操作在亚洲、南美洲和南非的城镇也渐成趋势。"全包式"的米饭料理使得工作日餐食的准备工作变得简单。新晋移民群体也越来越追求便捷，他们现在只在特定的周末和节假日烹饪传统家宴了。

并不意外的是，速冻即食饭大都由那些拥有悠久食米传统的国家生产。贸易市场的竞争促使泰国把冷冻速食产品运到美国来卖。英国则打算把（英国版本的）咖喱鸡等冷冻印度饭食卖回到印度的超市。这些"增值"过后的传统出口商品拓展了速食米饭的世界版图。无论你在哪里，它都可能出现在你的餐桌上。

一些经典的美国米制品

2006 年，美国人平均要花费其收入的 13% 来满足味蕾。而这项花费中的 40%，都用在下馆子（横向对比一下，吃路边摊和去餐厅的花费占非洲家庭和亚洲家庭饮食开销的 15%~50%）。随着城市的快速发展，出现了一些小公寓，这种公寓只有迷你厨房，或者一层有一个公共厨房（例如在俄勒冈州的波特兰市），甚至在香港和曼谷有根本不设厨房的房型，这导致人们对于开袋即食和可加热速食的食品需求飞速上涨。这些已经完全处理好的食物往往只需用加热板或者微波炉加热一下即可。在速食主义的进程中，米饭的地位是不容小觑的。

储藏室或者食物柜本来是用于囤放蔬菜、罐头、腌菜、咸肉或风干肉、鱼和水果等过冬物资的地方。随着城市的兴起，"超市"取代了"储藏室"。不像过去，很多冬储的食材需要打猎或采集获得，现在都可以购买到。在城市里厨房小、储藏空间小的公寓很常见，而郊区的房子有更大的厨房和储藏空间（存放方便食品和冷冻食品的好地方）。

第二次世界大战以后，越来越多的女性涌入职场。从 1948 年到 1985 年，女性在劳动力中的比例从 29% 上升到了 45%。由于职业女性回家做饭的时间缩短了，在主打女性市场的超市里，宣传的重点变成了如何快捷地做好一顿饭。

1904 年，"叱咤礼炮"膨化大米在圣路易斯世博会闪亮登场。引起轰动的正是身为植物学家和谷物产品发明者的

桂格膨化大米片，一种很受欢迎的早餐谷物。

亚历山大·皮尔斯·安德森。他知道预加工米中的水分在高温高压下会迅速膨胀、变成蒸汽，这样米粒就从里面被炸开了。第一个把这种膨化大米投入商业化生产的美国品牌是 1927 年成立的"脆米花"谷物。它早期投放的一个面向儿童听众的电台广告宣称，这种米花在牛奶碗里会"啪啪作响"并且保持酥脆。事实上，早在"脆米花"诞生之前，膨化大米就在印度生产多年了，而且还是当地很多小吃的灵魂食材之一。例如玛姆拉（mamra），就是用咖喱叶、盐、糖、姜黄、咖喱酱和大蒜调味，将膨化大米炒至焦香的小吃。类似的还有波哈（poho chevda），只要把膨化大米或薄米片的调味换成茴香、芝麻、盐、糖、油和大蒜即可。

　　1941 年，美国人又开发出用融化的棉花糖和黄油包裹膨化大米的火爆甜点 / 小吃——米花糖。如今，在报刊亭和超市里很容易就能找到开袋即食、独立包装的米花糖。这种商品化生产包装的小零食在美国军队里都是抢手货。

米花糖也在英国、欧洲大陆、加拿人和澳大利亚售卖。

1942 年，德国化学家埃里克·赫森劳布将预煮烹饪法的专利卖给了"本大叔种植园大米"水稻的卖家戈登·哈维尔。预煮法包括打磨之前的预处理，它可以将米糠中 80% 的营养素挤进米粒里，让精米更具营养价值。美国军队成了哈维尔的主要客户。到了 1944 年，这个品牌每年生产的 2 万 ~ 3 万吨大米被全数购置为军粮。第二次世界大战后，本大叔种植园大米得以相继推广到美国、加拿大、澳大利亚和英国；直到今天，它还是美国卖得最好的大米之一。在市面上，你很容易就能找到各种调味和半成品类的"本大叔"米制品。同样，这种预熟和干燥处理后的大米也以各种形式出现在好几代印度人的餐桌上；这又是一个为迎合新

膨化大米谷物片：要喂饱自己哦！

市场而重现的旧点子。

1958 年，文斯·德多梅尼科将家传的米饭和细意大利面食谱投入市场，他的独家秘籍是先用黄油炒香食材，再放入鸡汤烹制。这种菜谱仿照了亚美尼亚传统，和烹饪抓饭的方式有异曲同工之妙（后者是先用油脂炒香大米，再在肉汤中焖熟）。而在德多梅尼科的速食版本中，预熟干燥后的大米和意面被放在一个盒子里，搭配脱水调料包来取代鸡汤。唯一需要添加的材料是水。因为这种食物一半是米一半是意面，德多梅尼科将它命名为"通心饭"（取自"通心粉"的英文后缀）。"通心饭"的第一则商业广告拍摄了其产地旧金山的标志性缆车，还留下了深入人心的招牌广告歌《旧金山美味》！

美式米糕是从亚洲和印度的传统版本演化而来的。光

米制零食：入口香脆，五颜六色。

是中国和日本就有无数种米糕和米饼：软的、硬的，薄的、厚的，甜口的、咸口的……应有尽有。到底是松软如印度米糕，还是酥脆如日本仙贝，则要取决于时间、温度、大米种类等因素的综合作用了。淀粉含量高的大米和着水能做出卷红豆馅吃的软饼，而比较稀的米糊做出来的就是可丽饼或者酥脆的米饼干。另一种大受欢迎的日式米食就是麻薯了。蒸熟的糯米被放进小蛋糕模具，加入豆馅或者冰激凌，最后压制成型。麻薯几乎在全日本的冷柜里都能找到，在美国的日本人聚居区也很常见。

梅屋米饼公司是美国第一个米饼生产商，这家公司由日本的浜野兄弟创立于1924年，目的在于服务自己所在的社区。虽然"二战"期间这两位创始人被扣押入美国监禁营，但在战后这个公司的经营状况有所改善，逐渐成为全美国米糕的供货商。在美国，糙米、精米和混合米果也因为低卡高纤的特点（尤其是糙米制造的那种）广受好评，有切达奶酪、酱油、芝麻和焦糖等不同口味。

1949年，通用食品公司推广了阿富汗人阿图乌拉·欧载–杜兰尼发明的包装熟饭。这个产品的衍生品不计其数，那种只需一分钟就能做好的单人微波杯米饭就是其中之一。这种带包装和网络食谱的速食充分反映了美国人口的移民特质。从咖喱饭到墨西哥番茄辣酱饭、亚洲鸡肉饭或是希腊米沙拉，包装熟饭既能帮你还原家乡口味，也可以让你尝试异国美食。

方便面也是上班族、学生、家长和老年人市场的主打杯装微波速食，因为这是一种快速、简单、"不用刷碗"的食物。面条或者米粉先被炒熟再被烘干，然后连同脱水蔬菜和调料包一起装进聚苯乙烯泡沫塑料杯碗里。吃的时候只要加入沸水，等待一两分钟即可。方便面于1958年由安藤百福发明，这种杯装热汤面还在日本最受欢迎食品发明的评选里拔得头筹。

餐　厅

不论小餐馆还是大饭店，米饭必不可少。然而，这并非亘古不变的传统。在美国与欧洲，随着以大米为主要饮食的国家的移民人口的增加，迎合他们以及我们口味的餐厅数量也不断增加。食客们并不总是在乎菜肴是否"正宗"，但无论多么牵强，拿"地道的美味"作为餐馆的卖点总是个不错的伎俩。

19世纪50年代，中国劳工在加利福尼亚州的"中国城"开了几家简单的小餐厅供同胞们享用家乡风味。一碗堆成小山模样的白米饭衬托了猪肉、蔬菜、豆腐和豆豉、酱油、蚝油以及葱、姜、蒜、芝麻油这些调味品烹制的美味。随着非华裔客人越来越多，菜单也进一步扩充。虽然华人会暗自分出真正的中国菜和外国人吃的中国菜，但食客对于一碗白米饭的渴望总是共通的！

我们现在钟爱的"炒饭"，最早是后厨处理剩饭的一种方法，是不会出现在菜单里的。然而，在五花八门的伪装

清爽、缤纷、低卡的米纸春卷是越南人的解暑小食。

下，各种口味的炒饭（猪肉、虾、豆腐等）成为粤菜甚至其他菜系餐馆中的主食。现在甚至有只需用微波炉热热就能吃的盒装速食炒饭。

20 世纪 70 年代来到纽约州、加利福尼亚州和得克萨斯州的亚洲移民，其文化背景往往可以从他们制作的米食中反映出来。在圣何塞和休斯敦都有正宗的越南汤河粉售卖；而洛杉矶和纽约城的餐厅里，人们吃着杧果椰浆糯米饭；在纽约的皇后区和布鲁克林区的唐人街，可以找到福建红糟鸡。你还能从不同的米食中辨认出印度尼西亚、韩国、新加坡、中国和马来西亚等其他地域的文化。

据民间传说，1574 年，意大利已经出现烩饭。一个在米兰主教堂工作的玻璃匠人尝试用藏红花为他烧制的玻璃制品添加一点鲜亮的黄色。有趣的是，不仅玻璃制品成功上色了，他还在婚宴上用藏红花给牛骨髓和米饭上了颜色，宾客们发现这道菜竟然很美味。这种菜肴后来发展为米兰烩饭（risotto alla Milanese）。烩饭传统做法的第一步是用黄油或者其他炒菜油烹饪高淀粉含量的短粒大米，让每一粒米都被油脂包裹。然后缓缓加入热汤或者热水，一定要持续搅拌米粒，让淀粉物质慢慢渗出，这是使米粒筋道的同时汤汁呈奶油质地的关键一步。黄油和帕玛森干酪碎在食用烩饭前撒上即可。如果一个意大利餐馆提供烩饭的话，菜单上通常会附上一则温馨提示："烩饭大概需要25分钟制作。"除了经典的米兰烩饭，现在的菜单上有更多花样可供选择。人们想出了更快捷的制作方法，这也让非传统配料变成了常见搭配，比如豆腐做的素食烩饭和无乳制品烩饭（用米露和坚果泥使其呈现奶油质地）。你现在甚至能买到盒装的烩饭。

　　海鲜饭的故乡是西班牙的巴伦西亚。海鲜饭的英文"paella"借用了西班牙词语"por ella"的发音，后者的意思是"为她"（据传说，这道饭是一位爱人做给他未婚妻吃的美食）。而另一个不怎么浪漫的版本说，"paella"指的是做这种海鲜饭必备的双把圆形浅口平底锅，它的西班牙语单词为"paellera"。海鲜饭最早是一种混合了大米、蔬菜、兔肉、蜗牛的杂烩，工人们在田野的明火上做好后常常就端着平底锅直接吃。当锅还很热的时候，底部的米饭会变成一层焦褐色、脆脆的锅巴。在美国，巴伦西亚风味的西班牙海鲜饭里一般会用短粒或者中粒的精米搭配鸡肉、辣味香肠、虾、贝类、蔬菜、各种调味品以及藏红花。你猜得没错，海鲜饭也有盒装或者冷冻的版本。

手抓饭起初是贵族独享的美味，在伊朗、阿富汗和印度被称为"pulao"。品质最好的手抓饭是用放了几年的印度香米制作的，因此价格十分昂贵。每个手抓饭厨师的目标都是让一捧饭中的每一粒米都芬芳四溢。做手抓饭的第一步是洗米，要浸泡并清洗大米直到淘米水变得澄清，看不到淀粉沉淀。捣碎的青豆蔻荚、茴香籽和丁香等混合香料在酥油中与焦糖渍洋葱相遇。接着把米倒入，让每粒米都裹上富油脂的调味料。加入水或者汤汁，文火焖至米饭上出现小蒸汽孔。接着盖上盖子再蒸 20~30 分钟。在锅内米饭表面放上一条干毛巾，有助于吸收多余的水蒸气，保持米饭粒粒分明，防止粘连。阿富汗人和伊朗人通常会在做熟的米饭中间倒入一点酥油，让它慢慢渗到整个锅的底部。再蒸 30 分钟左右，锅底就会出现香脆焦黄的锅巴了。

秋葵浓汤的起源是个争议不断的话题，然而人们还是在几件事上达成了共识：这虽然听上去有点矛盾，但的确是这道美食历史的一部分。秋葵浓汤是一道以秋葵命名的、克里奥尔或者卡真菜系（两者均有争议）中重要的汤泡饭。而法国殖民者也跟这道菜有点关系，据说他们教会了当地人制作油面酱——一种让汤汁浓稠的油脂面粉混合物。受到北美乔克托部落的影响，有的秋葵浓汤会用檫树粉末来增稠。卡真人的秋葵浓汤里是加小龙虾的。要注意区分，加辣香肠可能是卡真的传统，但加烟熏肠则是德国的发明。是否放西红柿就要看厨师的喜好和他继承的传统了。

街头美食

街头小吃可以算是最方便的食物了。原先出现在路边摊、餐车、露天集市或者节日庆典的小吃，逐渐演变为中国、印度、泰国和墨西哥餐馆林立的美食广场里的"民族"快餐。任何可以卷着吃、夹着吃或其他吃起来方便的食品都是街头小吃的理想选择。在中国昆明，你可以吃到糯米纸卷着、外脆内甜的草莓冰糖葫芦；在印度孟买，你可以捧着羊肉香饭大快朵颐。或许你可以试试纽约韩国城卖的紫菜包饭（kimbap）——烤海苔片包着蒸好的米饭，里面裹着各式泡菜或者金枪鱼；或者里约热内卢的炸米球（bolinhos de arroz）——沙丁鱼和奶酪馅的碎米丸子；也可以尝尝新奥尔良的大米煎饼（calas）——一种裹着糖霜的油炸米糕。

有的街头小吃不只出自一人之手。像越南汤河粉、美式墨西哥烤牛肉豇豆饭、冈比亚亚萨炒鸡配饭这些食物，它们源自越南、美国得克萨斯州和冈比亚，随后也被移民带到了纽约州、新墨西哥州和加利福尼亚州。

全家出门觅食的话，吃路边摊算是一种实惠的选择了。在中国南方，竹制的小蒸笼像金字塔一样层层叠叠地垒着，食客们急不可耐地剥开蕉叶，品尝热气腾腾的蘑菇糯米饭。菲律宾的吕宋岛上，包在椰树叶里蒸熟的茉莉香米充满了椰子的香气，当地人对此十分买账。印度南部流行的早餐则是口感蓬松、像空气一样轻盈的蒸米豆糕。不要忘了还有多萨斯，一种用发酵米和豆子做饼皮、包裹着咖喱蔬菜、芥末籽十豆或者其他各式小料的"印式可丽饼"。而在印度尼西亚，明星街头小吃当属用马拉盏（虾酱）调味、鸡蛋碎点缀的印尼炒饭。

路边摊和流动餐车数不胜数，为繁忙的社区、美食节、农贸市场和集市提供现做的饭食。这些移动的盛宴为来往的人们奉上奶油咖喱菜配调味印度香米饭或者有泡菜和猪肉的韩式炒饭。不同于大多数情形，洛杉矶那辆掀起"餐车运动"的韩式烤肉塔可餐车最近变成了一家实体美食店。店里都提供什么呢？用肉丸、猪五花或者豆腐做主菜，搭配泡菜的韩式泡饭。

寿司：一个特例

街头美食有时精致且昂贵，有时又便宜且简单。而寿司则两者皆有。

据说约在公元 2 世纪，在靠近湄公河的中国内陆地区、老挝和泰国北部就出现了把盐渍鱼类放在两层熟米饭之间的做法。放置这些原料的容器会被密封很长一段时间，其间盐和米饭进行发酵，达到保存鱼肉的效果，最后弃米而独食鱼肉。然而，这种食物具有强烈的气味，对它痴迷的食客并不多（这有点像蓝纹奶酪）。长时间的制作和不低的成本也让它成为富人的美食。当然，如果是住在河边的人，获得鱼肉就相对简单了。

在公元 7 世纪的时候，这种类似寿司的形式就传到了中国和日本。然而中国人对这种食物不太感兴趣，日本人却很习惯吃米饭和鱼。这种鱼米组合在日本大受欢迎、需求高涨，因此食物做好后的储存时间就缩短了。日本人转而在确保鱼肉本身没有受到细菌污染的同时，保留了米饭本身可口的味道。公元718年，寿司甚至成为一种政府认可的赋税形式。到了17世纪初，寿司米加入了革命性的调味品——大米酿成的"米酒醋"（事实上它跟酒没什么关系）。用米酒醋调味的米饭可以和鱼搭配在一起吃，因为米酒醋不仅像发酵大米的味道，同时也有一些保鲜的功效，使得同时食用新鲜的鱼类和微酸的米饭成为可能。寿司也变得更便宜和平民化，由贵族的专享变为经常光顾街市的工人的午餐。简单的米饭和鱼的组合，精心制作的寿司便当盒，加上各式腌菜和其他配料都可随时供人们享用。

寿司

第二次世界大战后，日本的寿司食档或者餐车逐渐转为更符合卫生标准的室内餐厅。随着日式餐厅的兴盛，寿司开始了它的环球旅行，不断适应着新的环境，发展出加州卷、反卷和糙米寿司这样的新形式。顶尖的寿司师傅会特别在意自己使用的大米品种，有些甚至会亲自打磨大米（一般用顶级的短粒米或中粒米）。寿司学徒也正是从如何准备大米开始学起的。

你在高端餐厅或者公司食堂都能吃到现做的寿司，但超市或者小商店一般只卖包装好的。后者大都来自工厂，模具机器会将米饭塑成一个个椭圆形或卷状，熟练的工人再放上切片的鱼肉。速冻的养殖鱼和品质一般的米饭大大降低了寿司的价格，让它从有钱人的珍馐变为大众的美味。因此，寿司也在世界范围内普及。有的寿司店在环形吧台座中间设置了传送带，客人可以从回转台上一小盘一小盘地取自己想吃的食物。

寿司"世界风暴"最典型的例子大概就是拥有不可计数的寿司店的巴西圣保罗了。最早在 1908 年，日本人作为咖啡种植园的劳工第一次踏上这片南美土地。自此，移民人口不断增加，现在这里已经成为日本在国外最大的民族聚居地。在这里，人们每月大概会吃掉 1700 万份寿司。和日本差不多，寿司在巴西也是首先作为富人的食物引进的。随着鱼米之味的风靡，米饭和鱼肉的自动化生产把价格拉了下来。现在，连小餐厅和沙拉柜台也会提供寿司了。反映巴西自身历史背景的一个变化是，除了寿司以外的其他日本食物，人们更喜爱用长粒米搭配。除亚洲以外最大的大米消费国之一就是巴西——平均每人每年要吃掉 40 千克大米。他们制作的寿司也经常用到杧果、草莓、生牛肉这样的巴西食材来体现当地的口味。

清　酒

　　发酵大米和水的混合制品在中国和韩国已经存在了几千年，而我们最熟悉的日本清酒是在 2500 年前出现的。人们用霉菌、天然酵母甚至更原始的口嚼法将糙米在水中发酵成稀粥的状态，然后再进行过滤，最后得到了呈淡棕色、有点浑浊的萃取液。公元 689 年，在中国人的帮助下，日本皇宫的新酿酒坊培育出可以提高酒精含量的发酵菌。在神道教寺院凭借自己的力量成为合法的酿酒厂时，饮用清酒的仪式就得到了巩固。在之后的四百年间，酿酒成了一桩生意，京都和神户是主要的酿造中心，政府也开始对此征税。在 16 世纪末，刚刚出现的精米也开始被用来酿酒，这也意味着制成的清酒可浑浊、可澄澈。等到 19 世纪，清酒变得像苏格兰单一麦芽威士忌、红酒和奶酪一样，受年份、大米品种、土壤、气候和水源等产地条件的影响，孕育不同的风味，消费者也会因为自己的口味对酒有不同的偏好。在大米短缺的时期（例如第二次世界大战后），人们会用少量的米和蒸馏后的廉价酒生产出品质很低的清酒。有趣的是，第二次世界大战后，威士忌、红酒和啤酒在日

清酒

本流行起来，清酒反而在欧洲、南美洲、澳大利亚和美国
受到欢迎。

在 1885 年的时候，清酒同糖料种植园的日本劳工一起
来到了夏威夷。因为不断上涨的进口关税，一些日本的酒
坊选择在美国本土设厂。最开始为了提高销量，这些酒坊
主要做低品质的清酒。在喝的时候，酒要拿去温一下，以
掩盖低劣的品质。另一个不同于日本饮食文化的习惯，是
用清酒搭配寿司。不仅传统的寿司礼仪明确规定不准饮酒
（"好上加好、锦上添花"在一些习俗中是备受鄙夷的），而
且高品质的清酒应该冷藏饮用，用古董饮具呈上，还要为
品鉴者提供年份和标签信息。

电饭煲

做米饭的方法通常有焖、煮（像煮意面一样）、隔水蒸，

或者混合使用这些方式。然而，任何一种方法做出来的米饭都有可能不甚完美，特别是在锅里煮饭，需要一边照看柴火和火势的大小，一边旋转锅让米饭均匀受热。诚然，可靠的煤气灶已经出现一个多世纪，但电饭煲的发明彻底革新了日本的米饭烹饪，也影响了整个世界。

第一台自动电饭煲是在 1955 年由东芝制造的。可以全天保温的电饭煲在 1960 年诞生之后广受欢迎，寿司店也搭上了这趟顺风车。1979 年，电子科技也让前一天晚上定时、次日早上吃上热米饭的幻想变为现实。松下在 1988 年开发出电磁加热电饭煲。虽然定价更高，但如今松下已经占据了电饭煲市场份额的一半以上，可谓后来居上。这个产品的好处在于，不需要提前泡米，并且结束时会有轻微的搅动（帮助多余的蒸汽挥发掉），米饭口感会更加稳定。2003年，松下又发明出一种使用极热蒸汽的电饭煲——这样做

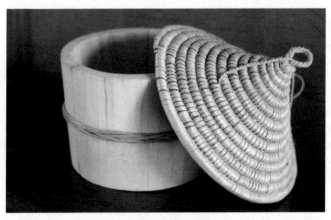

中国云南的蒸饭桶：我把它买下带回了纽约，真的很好用。

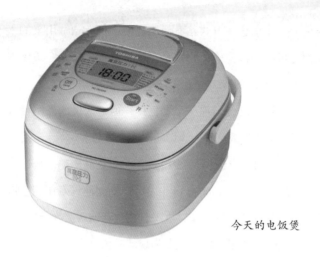

今天的电饭煲

出来的米饭会更香。

　　松下发现，欧洲人和美国人都对电饭煲饶有兴趣，因此根据顾客的偏好生产了特别定制的款型。在美国，电饭煲上被加了一个篮子，用来蒸蔬菜。盖子是透明的，这样就能查看里面的食物是否熟了。电饭煲品种的选择也要根据不同的大米品种进行调整，大米是否预煮过或者你想要蒸熟还是煮熟的口感，都是不一样的。就像大米一样，电饭煲也需要适应食客的需求。

艺术、礼仪和象征

幸运就像飞进嘴的糯米点心。

——日本谚语

很多起源神话、习俗、礼仪、语言和观念都与水稻息息相关。在人们移民到别处，或者运用现代科技之后，一些传统发展成了新的法则。杀虫剂取代了虔诚的祈祷，速生水稻也胜过了传统的品种。在机器取代人力之后，现代的水稻生产技术也造成了传统的丢失和失业率的上升。

依靠补助金支持，亚洲的研究中心一直企图通过直接播种、机械化水稻插秧、除草剂筛选和机器脱粒等方式来减少劳动力，这些尝试着实令人担忧。这主要是个关乎时间点的问题。随着经济的发展，从事农业生产的人口已经到达一个从过剩到短缺的转折点……这一显著趋势的标志是，在很多地区，那些已经使用了数个世纪的水稻种植技术和做法正在逐渐消失。负责耕地的不再是水牛，而是拖拉机；直接播种代替了育苗移植；除草剂省去了手动除草的麻烦；机械化的脱壳设备让农民不用在稻田间拿手搓谷子了……年轻人不再把水稻种植纳入考虑范围，而那些被留下照看稻田的人则不得不适应新的技术去减少劳动负担、提高耕种效率。

尽管国际水稻研究所在 2001 年的"21 世纪水稻研究与

生产"报告中指出上述的变化，但一些关于稻米的古老传统和习俗依然备受重视、延续至今。在稻田里，你还可以看到人们安置的神龛；在庆贺丰收的宗教仪式上，神龛也很常见。河内的水上木偶剧团最著名的表演就是模仿水稻种植的盛景和丰收的文化象征。在泰国，那些信仰稻谷女神梅普素的人会在丰收时小心翼翼地砍下秸秆，以免冒犯到她。而在印度，人们会将篮子里的稻谷撒在新婚夫妇的头上。美国的婚礼上也会撒稻谷（如果是室外婚礼，现在已改成撒鸟食，因为鸟类无法消化生米）。也就是说，一些起源神话是有相通之处的；如果你知道水稻是如何以复杂的方式从一个族群传到另一个族群，从一块大陆到达另一块大陆，这件事也就没什么稀奇的了。

稻田里的稻米神龛

然而，并非所有的水稻意象都是积极的。稻米和水稻种植也象征着负面的种族和性别分化。

起源故事与神话：神明和水稻

主食通常都有很多个版本的起源故事。一般来说，神明都有赐予或者没收基础食物的能力：红酒、啤酒、玉米、巧克力、麦子和水稻都涵盖在内。为了确保五谷丰登，人们往往会去神龛前或者寺庙里参拜，向动物神或者人形男神女神奉上祭品。一场稻米的丰收，不仅保证了人们当下的存活，还能帮助抵抗未来可能遇到的粮荒。

一些神是有报复心的，也能感知悔恨。关于爪哇的蒂斯娜瓦蒂的传说就是个例子。蒂斯娜瓦蒂本是天神的女儿，却爱上了凡人雅加苏达纳。蒂斯娜瓦蒂的父亲不允许女儿与人类有牵系。作为惩罚，他将蒂斯娜瓦蒂变成了一株水稻。后来父亲可怜她凡间的爱人，于是把他也变成一株水稻，栽在爱人身侧。他们重新相会的日子正是丰收节，这个结局也象征了矢志不渝的爱情终会战胜不可靠的一时冲动。

马尔科姆·格拉德威尔在 2008 年出版的《异类：不一样的成功启示录》一书中指出，水稻丰收的三要素——专注的投入、努力工作的意愿、精心安排的团队合作，可能与数学推理有着密不可分的关系。虽然这是一种成见（这也是我会在本章写这段内容的原因），但他举了中国南方的例子：稻田面积一般很小，但全家人都要参与耕种，照顾水稻的方方面面。确定播种时间、测算面积、培土、定水位、挖水塘、除草、灌溉和其他繁重的水稻种植工作都仍然需要人力。他们几乎每天都需要做决策，而且会涉及算术和分数。再加上中国年轻人更强的数字记忆能力（相较于美国人）和工作时的不知疲倦（不论是在田里还是在教室或者家中），亚洲人称霸美国数学课堂的现象便不难理解了。水稻种植和超强的数学能力成为新加坡、中国、韩国和日本共通的文化基因。

也有些神是很慷慨的。比如在印度的泰米尔纳德邦，稻米女神叫作波尼亚曼。这个名字是此地水稻品种"Ponni"和泰米尔语里女神一词"amman"构成的复合词。这个区域经常遭遇洪水的侵袭，导致稻田被冲毁。于是，农民就在稻田间立起了波尼亚曼的雕像，对她虔诚参拜，女神也应了他们的祈求，让洪水不再泛滥。

根据中国一个有关大米起源的传说，洪水冲垮田地之后，有个人在山间避难。一条小狗从他身侧跑过，尾巴上吊着一株稻穗（稻穗就是水稻分蘖节上长颖花的地方，成熟后就是谷粒）。小狗经过之处都有种子掉落，之后迅速长出水稻幼苗。

还有一个神话是说，一位名叫胡吉尔的猎人有五个儿子。他给了每人一个麻袋，让他们选择麻、麦、黍、菽、稻这五谷中的一种将袋子装满。选择了稻米的儿子名叫帕迪——这就是水稻的起源，英文里水稻田"paddy"一词正是取了他名字的谐音来纪念这位伟大的先祖。

菲律宾的布鲁尔塑像：请在碗中装满大米，让我的家人饱餐一顿。

　　在菲律宾伊富高族的水稻梯田里，男女成对出现的布鲁尔雕塑正是族人的水稻神。相传神明修米德希德用菲律宾国树紫檀树的树干雕刻了四尊人像。这些小雕像顺着河水漂流而下，经过稻田和谷仓便会繁衍出更多的小人儿，留在那里守护一方水土。

　　稻荷是日本稻谷之神，相传就是他种下了第一株水稻。曾有一条蛇镇守一捆水稻，稻荷神的神使狐狸则受命在大地上播下水稻种子。日本家庭常会在家中置放稻荷神龛，日本神社和佛祠也会颂扬稻荷。这位脚踏两袋水稻、身骑狐狸的神仙也被称为"财神"。关于这个主题流传的故事还有很多其他的版本。

庆典：水稻节日的范本

在巴厘岛，丰收节祭奠的稻米女神丝莉有三种化身：其一是我们之前在爪哇文化中提及的蒂斯娜瓦蒂，她是水稻的创造者；第二位就是水稻的保护者湄公母（也是泰国的水稻女神）；最后，丰收的水稻会幻化成凡间女子的模样。关于水稻的由来，有这样一种说法：天神巴塔拉·古鲁得到了一颗蛋状的宝石。打开后，一位妙龄女子现身，神给她赐名为蒂斯娜瓦蒂。然而这位佳人很快香消玉殒，这让众人悲痛不已。在她的葬礼之后，国王在森林里骑马时发现在蒂斯娜瓦蒂的墓旁，出现了一束耀眼的光芒。他连忙靠近，竟然看见女人的头顶生出一棵椰子树，手上托着香蕉树，牙齿变成了玉米，私处长出水稻。

在节日期间，村子里会重新在墙壁上彩绘，到处装点彩旗。田野里满是沾着米粉的雕像，以此纪念稻米女神丝莉。象征着母性精神的稻草人会立在稻田旁，确保岁稔年丰。米仓是先祖神明的居所，因此总是被建成小房子的样式，里面储存的大米足够他们享用到下一个丰收季。米仓里也会放上一些用于庆典祭祀的物件，一般是家中年长的女人做的。庆典上还会吃烤鸭和猪肉、印尼炒饭、米制甜食和饺子。

泰米尔纳德邦每年都会举办为期四天的庞格尔丰收节来感谢阳光、雨露、牲畜和谷物（主要是稻米），还有甘蔗和姜黄。在第一天的黄昏时刻，当地人会燃起篝火，将老

用稻草做成的泰国
稻米女神丝莉

旧无用的东西扔入火焰中。他们会清扫房屋，并用精致的
五彩米粉画装饰家门口的土地。在收割新一茬水稻前，农
民会用檀香油涂抹镰刀和犁。第二天是祭拜太阳神苏里亚
的日子。人们会向他供奉在砂锅中用牛奶煮熟、以姜黄和
甘蔗装饰的米饭。在锅煮开后，大家也会分食一些米饭。
在第三天的时候，人们会洗干净水牛和奶牛，给它们戴上
花环并参拜，这是为了感谢它们犁地的苦劳。农民也会喂
它们吃用牛奶煮熟的米饭。到了第四天，年轻女人们要准
备好小饭团放到田里供鸟儿吃。庞格尔节也是适合结婚的

庞格尔大米画：五颜六色的米粉设计艺术为一年一度的印度庞格尔节预热。

吉利日子，因为丰富的物产可以给婚礼提供充足的大米和其他必需品。

在日本大阪，6 月的插秧节依然遵循旧礼进行。稻田还是要由公牛犁好。在大城市里，水稻已经种到了摩天大楼的屋顶上。但日本人认为，水稻的种子中有生长之灵，因此必须直接由大地母亲孕育，所以节日上别致壮观的歌舞表演正是为了唤醒沉睡的种子。仪式上，女人们戴着插满鲜花的斗笠，边唱边跳，全副武装的武士从她们身边列队前进，庆典的高潮是由 150 名年轻女子表演的"住吉踊"①。秋日的大丰收便是神灵对人们在庆典上的祷告

———————————————————
① 日本的一种民间舞蹈，传承自大阪住吉神社。

做出的回应。10 月一收割完，人们就要拿一些稻米来供奉神祇。每年的 11 月 23 日，还会有一个庆典来感谢当年的收成。

在中国，腊八节、春节（中国农历新年）、端午节和中秋节都是有关稻谷的节日，此类节庆日在日常生活中还有很多，庆祝的内容包括新米的到来、丰收、夏至与冬至和我们即将说到的媒妁婚约。在贵州苗族的姊妹节中，姊妹饭既是一种美食，也是极具象征意义的文化符号。年轻的女孩子会用摘下的绿叶、鲜花和青草，制成天然的着色水，将大米浸泡几天，最后做出颜色鲜艳的米饭。她们会在饭中放上信物，再把姊妹饭交与看上的男孩子。不同的米饭

选自一个记载婆罗门一生的图册，图中画的是 1820 年左右小孩出生后 11 天取名时的庆祝活动。家中的妇女在后排享用食物，前面就是男人和他们请来的婆罗门。每个人都以大芭蕉叶为托盘，上面堆着米饭、香料卷饼、蔬菜、茶和几个杯子。

颜色和信物有着不同的意义。例如，红色代表了女子所居的村庄繁荣昌盛。如果饭里有棉花，就代表女生想赶紧结婚；如果是蒜头，则意思正相反。

另外一个年代更近的例子来自中国西南部云南省的蒙自市，庆典的食物是过桥米线。相传一位男子在离家很远、安静偏僻的南湖岛上准备科举考试，他的妻子每天中午都给他送饭，但每次等她过了桥，汤水就冷掉了。有一天她煮了锅鸡汤，到丈夫书房的时候，汤还是滚烫的，因为鸡汤上面漂浮的油脂就像密封层，在长途跋涉中依然能保持汤的温度。因此，鸡汤虽然不是主要食物，却成了这顿饭的基础。米线以及切得很细的肉、鱼和蔬菜等生食材会被用来搭配鸡汤。吃之前，把这些食材丢进滚烫的汤里，这

中国云南的过桥米线：一个为婚姻和学业献身的故事。

样就有一碗新鲜烹制的热饭吃了。

在朝鲜和韩国，面条（寓意为长寿）和饺子常在生日宴上出现。蒸年糕会用板栗、蜂蜜、红枣、高粱米和艾蒿调味。特别是在韩国，用艾叶做的年糕常被认为有药用价值。这些年糕也是韩国端午祭必不可少的食物。

在李氏朝鲜初期（1390—1910，尤其在16—17世纪之间），随着儒学的传播，节日仪式和节气庆典兴盛起来。鼓励人们通过食用年糕汤、松针蒸年糕、甜米露等食物，实现天人合一的理念。韩国人会在新年喝年糕汤；在小孩百天的时候，要给他一块代表天真纯洁的蒸年糕；在孩子一岁生日时，会准备像彩虹一样色彩分明的分层年糕，预示着未来的拼搏。

在很多西非的庆典上，大米也是很重要的。虽然妇女承担着大多数田间的劳作，但是各种典礼仪式上戴着面具和头饰模仿动物和鸟类的假面集会却是由男人完成的。以米为主食的宴会证实了大米与女性生育能力之间的联系。细致的木雕和用芦苇等草叶编织的簸箕也是仪式重要的组成部分。时至今日，非洲人还会做这样的草篮子，而且在美国的南卡罗来纳州周边的低地岛屿上也能见到它们的身影。

为了促进丰收，利比亚和科特迪瓦的丹族妇女会拿着刻有动物和人形的木质长柄饭勺翩翩起舞；在马里，班巴拉族的男性会戴着羚羊角头饰，跳敬畏神灵的舞蹈。而在几内亚和利比亚的巴加族婚礼上，新娘会头顶一个篮子跳舞，周围的人则向篮子里掷大米和钱币作为礼物。

在美国，阿肯色州、路易斯安那州、南卡罗来纳州和得克萨斯州都有一年一度的大米节。大米节是庆祝丰收、敞开享用米制美食的日子，也彰显了大米产业对于当地的重要意义。在路易斯安那州的克劳利举办的国际稻米节涵盖花车游行、"米国王"和"米王后"的加冕仪式、米制美食会以及米食烹饪大赛等活动。从血肠（一种米制香肠）到什锦饭和秋葵浓汤，这里的米制美食可谓应有尽有。

10月的时候，一年一度庆祝丰收的得克萨斯大米节会在维尼镇举办。该节日以嘉年华、游行、家畜和长角牛比赛、马术表演、烧烤大赛、夜间街头舞蹈秀、米食烹饪比赛、选美比赛等活动为特色，同时结合了当地盛行的卡真风味的米制美食。经典的节日美食有饭团、秋葵浓汤、小龙虾焖饭（一般是颜色厚重的酱汁包裹的小龙虾或者其他海鲜，浇在米饭上）、蟹肉丸、血肠丸（香肠和米饭炸制的丸子）、漏斗蛋糕[①]和经典的牛仔炖锅——碎牛肉、蔬菜、豆类和西红柿做的一锅炖，最后再放入饼干、玉米面包或者米饭一起烘烤。

从1976年开始，每年的阿肯色大米节都会票选出一位阿肯色大米王后。各式各样的烹饪大赛也都在这个时候举行，常常吸引本地的大厨来一展身手。米饭甚至会与黄油和砂糖一同食用。

---

① 北美流行的一种油炸面糊食物，因制作时会将面糊经过漏斗滴入油锅而得名。

在西班牙巴伦西业南部的一个农业小镇苏埃卡，海鲜饭是当地大米节上的明星。这个节日会在每年九月如期而至，庆祝这个国家享誉世界的西班牙海鲜饭。本地和外国厨师会用当地著名的大米品种参加"国际海鲜饭大赛"，包括"邦巴米①"以及其他的特殊短粒米，其如同长粒米一样圆润，且吸水性好。传统的巴伦西亚海鲜饭会用到鸡肉、兔肉、蜗牛和一些绿叶菜，但每个小村子都有他们引以为傲的独门秘方。最出名的要数鸡肉和兔肉的版本，当然还有海鲜饭和双拼口味。西班牙渔民创造了海鲜饭，为了保留米饭和海鲜的原汁原味，两种食材一般会分开制作。这种饭一般会佐以蒜泥美乃滋。此外，放有甜菜根、乌贼、花椰菜和菠菜的砂锅焗饭也很受欢迎。

文化习俗和稻米传统

向新婚夫妇抛撒稻米是亚述人、希伯来人和埃及人的古老传统。在印度的婚礼上，当新娘和新郎围着篝火转圈时，新娘的兄弟要将未脱壳的谷粒倒入他们手里。夫妇二人再将这捧稻谷抛向火堆。大米不仅是印度新娘送给丈夫的第一顿食物，也是婴儿出生后吃的第一种固态食物。大米几乎是孕育生命的同义词。

在日本的婚礼上，新人会被送上印有仙鹤和乌龟图案

---

① 一种短粒米，主要种植在西班牙东部，常用于西班牙海鲜饭中。

的米糕，寓意长寿。红豆和大米会被当作婴儿的诞生礼物，而一场佛教的葬礼中会用到膨化大米，意为无法再次生长的稻谷。在韩国的葬礼上，已故之人的嘴中会放三勺米和一些钱币，来帮助其顺利转世。

语言、文化和艺术

在一些亚洲语言中，"米""食""餐""吃"几乎表达了相同的意思。"农业"和"大米"也常被用作同义词。大米在表达中的用法多得不计其数。

虽然圣经中没有明确提到米，但孔子和穆罕默德最爱的都是米食，后者喜欢用酥油（澄清的黄油）烹制的甜口米饭。佛祖曾是尽享荣华的悉达多王子，后来决定通过苦行修炼来获得顿悟。在数月里，他每日只食一粒米。有一天，一个小女孩给他带来了一份牛奶焖饭。他吃后恢复了精神气力，得以继续他的探索。据说佛祖的长袍上还绣有稻田图案，这种设计一直到今天还在使用。

克利须那神①的信徒们认为，根据神的旨意，食物被分为三大类，第一类包括米、牛奶和其他乳制品，都与美德相关。

很多谚语都是用大米来表达特殊的含义和感情的：

---

① 即黑大神，婆罗门教、印度教最重要的神祇之一。

莫让愠怒之徒洗餐碟，莫让饥饿之人守谷堆。

——柬埔寨谚语

一年之计，莫如树谷；十年之计，莫如树木；百年之计，莫如树人。——中国谚语

顾一钵之米，而失一桌之宴。——越南谚语

米应生于水，而终于酒。——意大利谚语

吃饭了吗？——泰国传统的问候方式

路易斯·阿姆斯特朗，这位伟大的美国爵士小号演奏家最爱的食物正是红豆饭。为了致敬这道新奥尔良标志性美食，他还特意在信件中留下"如红豆饭般温暖的你"。

大米象征主义的另一面

一些和米有关的用语里也根植了一些负面的种族刻板印象。在越南战争时期，南越妇女和非裔美国士兵生下的孩子有时会被形容成"煳米色"。同一时期，南越的动员海报上画着大米丰收的景象和一排举着刺刀的士兵来鼓舞士气。意思是，只要有充足的稻米作为军粮，士兵们就能安心在前线持续作战。

美籍华裔女作家兼教育家汤婷婷在《女勇士》（1975）一书中，也用了跟米有关的说法来描写其祖父对于女孩的看法："女孩离开家是件大好事。女人就像米中的蛆虫。养几只鹅都要比养女孩来得划算。"在菲律宾，"ampao"是一

种膨化大米美食的名称，但当用于贬义时表示一个人脑子不好使的意思。

在《婚礼的成员》（1946）中，卡森·麦卡勒斯描述了主人公对于大米的挚爱：

> 现在，豌豆饭是 F. 贾丝明最爱的食物了。她反复告诉身边人，当她躺在棺材里的时候，千万别忘了拿一盘豌豆饭在她鼻子前面挥一挥。如果她还有一息尚存，一定会坐直并大快朵颐；但如果她在豌豆饭面前都没什么反应，那她就是真的死了，他们可以钉上棺材了。

大米在下面这首菲律宾歌曲里是用来赞颂辛苦的体力劳动和苦工的：

> 种水稻很无聊，
> 从早到晚弯着腰，
> 不能站，不能坐，
> 一刻休息得不到。

神奇的是，第二次世界大战后，菲律宾的美国公立学校体系里竟然会响起这首民谣，但无论人们选择它的初衷有多好，看起来它已经失去了原本的意义。

文化与大米：以日本为例

大米对于日本社会的重要意义已经被广泛研究过了，它提供了一个窗口，让我们得以瞥见更为详细的象征文化价值。

通常认为，族群和谐、依赖和共识这些概念正是起源于水稻种植的过程中。早先，家庭是聚集劳力和技术的单位。水稻种植是劳动密集型工作，每个人都需要同时掌握精湛的技术。然而播种、修筑堤坝和分配水资源这样的工作，则需要区域内的家庭联系起来。所以大家把房子建在一起，团体间互相帮助，各家的土地都一起耕种。收获的时候也是一样的情形。共同决策和群体利益高于个人偏好。对于将世代共事的邻里来说，避免摩擦是很重要的事情。作为原始的水稻文化的一种特征，这种历史性的为群体协同而投身的精神一直延续到今天，塑造了日本人的集体意识。虽然只有很少的日本人还在从事水稻种植的工作，但在这片狭小的土地上，还有 1.24 亿国民仍然在日常生活中恪守着人与人和谐相处的原则。

在日本的语言文字中，也能找到对这些概念和价值观的体现。米饭在日常饮食中的首要地位反映在语言中。"ご飯"既可以表示"做熟的米饭"，也是"一餐"的意思。从此衍生出来的日语词还有早餐（朝ご飯）、午餐（昼ご飯）和晚餐（晚ご飯）。这样的语言信号清楚地表明了"无米不成席"。

水田艺术：用稻秆、禾苗和其他谷类再现的"名画"《神奈川冲浪里》。

　　另一个体现稻米与语言息息相关的标志，就是早年间日本当地人对自己国家的称呼——瑞穗国（长满水生根茎作物或水稻的土地）。有趣的是，日本人称美国为"米国"，从而表示该国富足。

　　从历史上来看，大米跟日本文化有着千丝万缕的联系。比如，在日本早期的历史上，天皇也是宗教领袖。根据神道教的教义，他主要负责与水稻种植相关的事务，也包括亲自制作清酒（米酒）和麻薯（年糕）。日本裕仁天皇曾在东京的皇家用地上管理一小块稻田，一直到他重病才作罢；即便那时，他还在担心着天气和收成。明仁天皇也延续着这一传统，为天下稻谷丰收祈福。很多加冕仪式都会用到大米或者米制品，这也强调了稻米与日本皇室和神道教密不可分的关系。

曾经，大米需要被好好看守，因为其代表着安定和繁荣，由此可见其重要的社会意义。"合"是一种衡量大米的计量单位，大米不仅被用作交易工具、硬通货和给武士的俸禄，还被用来度量财富。

其实日本生活中的方方面面都能提供很多例子，像民俗、节日、艺术和家庭礼仪都能作为这节概况的补充。水稻的所有部分都能为日本人充分利用。每年有大概32千克秸秆在被回收后用来编织榻榻米席子，很多日本家庭都用它来铺地板；米糠被用来制作脸部磨砂膏；米浆用来装订书页，它也是一种织物防染色材料，经常用在制作和服用的丝绸上。大米与日本文化密不可分。看到月亮时，美国人会谈到西方传说中的"月仙"或者联想到美女的脸庞，而日本人则会看到一只捣年糕的小兔子，这是他们家喻户晓的民间故事。

食

谱

食谱和烹饪技巧展现了历史长河中特定的时间和地点。古时候，大多数食谱都是口口相传的，后来才被人用笔记录下来。这些食谱先是在为贵族服务的厨师和神职人员中流传（无论是在罗马的阿比鸠斯[①]时代或者20世纪初的埃斯科菲耶[②]时代），后来在中产阶级中得到普及。到现在为止，近代阿拉伯文明贡献了最多的烹饪书籍，并且有实物留存至今。最古老的菜谱是用散文文体写的，需要极高的烹饪技术。

我在下面首先列出了两类食谱——手抓饭和秋葵浓汤——的例子，它们在很多国家都是大受欢迎的美食，也有不同的版本，此后的各种特色配方混搭体现了它们跨越文化的魅力。

肉末香料饭

查尔斯·佩里翻译的13世纪菜谱，选自中世纪的
阿拉伯烹饪书《菜肴之书》

做这道菜需要把肥肉切成中等大小的肉丁。熔化新鲜的羊尾油，把油渣丢掉，之后放入刚刚切好的肉，炒至焦褐色。稍稍撒一点盐和干香菜末在上面。倒入没过肉丁的水炖煮，记得要撇去浮沫。在收干汤汁、肉开始焖煮时就可以出锅，避免煳锅。这时放入足量的干香菜、茴香籽、

---

① 罗马帝国时期的一位公认的美食家。著有《论烹饪》等与烹饪相关的作品。
② 法国著名的餐馆经营者与美食家。

肉桂、研磨细腻的乳香脂和盐巴。调好味后，把已经脱去水分和油脂的肉丁从锅里盛出来，再在表面撒一些刚刚提到的香料。准备好一量杯的米和三量杯（或多半杯）的水，再熔化重量为肉碎重量三分之一的新鲜羊尾油。锅中注入清水，煮沸后加入熔化的油脂。然后放入乳香脂和整根的肉桂，煮至整锅沸腾。淘洗几次大米，用藏红花给它染色，然后倒入水中，注意不要搅动。盖上盖子煮米直至水烧开，米饭煮熟，开盖铺上刚刚准备好的肉。再用布盖住锅盖并把锅包好，防止空气进入。转小火，继续焖到没什么咕嘟咕嘟的声音就可以出锅了。有些人的做法比较简单，省去了拿藏红花给米饭染色的步骤。

查尔斯·佩里在脚注中写到，"aruzz"是米饭，而"mulfalfal"表示"烹饪得像胡椒粒一样"，也就是粒粒分明的意思。他还提到，这个菜名也有可能是受了波斯语"pulau"或者"pilaf"一词的影响。

葡萄干松仁手抓饭

手抓饭也能用碾碎的干小麦和大麦做，但大米仍是首选。克劳迪娅·罗登在《中东饮食新编》（2000）里写到大米是由印度传到波斯的，并由阿拉伯人传播到西南的西班牙，最南到西西里岛。白米饭在阿拉伯叫"roz"，在土耳其称"pilav"，在伊朗是"chelow"，加入其他配料后则叫作"polow"。手抓饭可以与炖菜一起端上桌，也可以做出造型，染上红色或者黄色，也可以和蔬菜、水果、坚果、肉类、鱼、奶油和牛奶一起烹制。它既能和其他菜一起上桌，也可以先后被端上桌，具体情况要看你在哪儿吃饭、和谁吃饭。很多个品种的长粒米都能做手抓饭，每种米也都有各自的拥护者，这都不足为奇。

我的手抓饭食谱是根据罗登的奥斯曼帝国宫廷经典食谱改编的。

2 个中等大小的洋葱，切碎

15 毫升（2 汤匙）芥花油

100 克（2/3 杯）烘焙过的松子

400 克（2 杯）长粒米

675 毫升（3 杯）鸡汤

1 茶匙磨碎的多香果

1 茶匙肉桂

1 茶匙葫芦巴籽

适量的盐和胡椒

3 茶匙金葡萄干

6 茶匙黄油，切成薄片

1 茶匙切碎的莳萝

在一口大号平底锅中将洋葱丁用油炒至金黄变软。加入松子和大米，用中火翻炒，直到米粒和松子都被油脂充分浸润，并开始变得焦脆。倒入鸡汤，放入多香果、肉桂、葫芦巴籽、盐、胡椒和金葡萄干搅拌。汤沸后盖上盖子转小火焖煮，继续煮20分钟，或者煮至大米变软。加入黄油和莳萝碎搅拌，趁热端上桌。

罗登注明：在土耳其版本的手抓饭中，热饭还可以加上调味腌制的鸡肝和切碎的莳萝。

秋葵浓汤

秋葵是一个词，一份菜谱，一道浓汤炖菜，一种社会写照。制作秋葵浓汤的配料可以说是一条连成烹饪世界的日界线，纷繁的种族文化交汇于此，诞生出美味又有趣的融合菜。以下两个版本的食谱都出自《卡罗来纳大米厨房：非洲纽带》（1992），书中还有对塞缪尔·G. 史通尼女士在1901年编撰的《卡罗来纳米食烹饪书》的简要介绍。

新奥尔良秋葵浓汤

将一只火鸡或者家鸡和一块新鲜牛肉一起切碎，把它们放入锅里，加入一点猪油、一个洋葱和足量的清水，将肉煮熟。煮软食材后，将100个牡蛎连同汁水一起倒进锅里。按照你的口味调一下咸淡。将汤搅拌至黏稠，撒上两勺黄樟叶粉末，即可出锅。

"配上米饭"一词并没有出现在文本里，但这是一般设定。秋葵也没出现，想必黄樟叶粉也可以发挥相似的增稠功能。

南方秋葵浓汤

将两个洋葱切片后翻炒，同时准备一只大小适中的鸡，斩好；把鸡肉也加入炒锅，与洋葱一起炒至表面焦黄。准备一夸脱①切好的秋葵和四个大番茄，把这些菜和鸡肉一起放进一个大炖锅里，注入没过食材的热水。汤汁煮至浓稠，用盐和红辣椒荚调味。这道菜一定要装盘，并和现煮的米饭一起食用。

米布丁

无论是在历史悠久还是后期发展的稻米文化中，都有米布丁的身影。它的形式多种多样，从最简单的又软又甜的米布丁到用炼乳做的西班牙牛奶饭，或者是埃斯科菲耶为了纪念欧仁妮皇后嫁给拿破仑大帝而制作的有精致外形的皇后米糕，后者需要用到香草卡仕达酱、鲜奶油和白兰地渍的水果。介于其间的，还有软的、韧的、可以切片的、能舀着吃的、白的、黑的和棕色的等各种版本的米布丁。从香草和柠檬皮，到小豆蔻、腰果、开心果和藏红花，都能给米布丁带来多变的口味。现在，牛奶、豆奶、米露和椰奶都可以用来制作米布丁。

---

① 夸脱是个容量单位，主要在英国、美国及爱尔兰使用。1夸脱在英国和美国代表的是不同的容量，而美国更有两种夸脱：干量夸脱及湿量夸脱。美制：1夸脱等于0.946升，即0.000946立方米。（干量单位，约等于1.10升）。英制：1夸脱等于1.1365升，即0.001136立方米。

美国殖民地风味米布丁

选自 J. M. 桑德森的《烹饪完全指南》（1846）

米布丁——取一杯卡罗来纳大米和七杯牛奶；把锅放水里隔水煮，保持水沸腾的状态直至锅里的液体变稠；再加糖，最后再撒上一盎司的甜杏仁碎。

印度炼乳米布丁

2 品脱[①]（1.1 升）全脂牛奶

2 茶匙长粒米，比如巴斯马蒂米

4 颗青豆蔻荚，稍微碾碎

10 粒无盐开心果

2 茶匙糖

装饰：

印度甜品金银箔装饰（蛋糕店中或者一些亚洲超市里能买到的可食用金箔或者银箔）、开心果碎、其他自选配料

将牛奶倒入一个厚壁的锅里，文火加热（也可以用微波炉将牛奶在罐子里直接预热，再将热牛奶倒入锅里，可

---

① 品脱是容积单位，主要于英国、美国及爱尔兰使用。1 品脱于英国和美国代表的是不同的容量，而美国更有两种品脱：干量品脱及湿量品脱。1 英制品脱 = 568.26125 毫升。1 美制湿量品脱 = 473.176473 毫升。1 美制干量品脱 = 550.61047 毫升。

以节约时间）。在牛奶中加入米饭和小豆蔻。

慢慢煮至沸腾，然后文火焖煮，时不时搅拌一下防止米饭粘在锅底。重复这样小火慢煮和搅拌的工序，直到牛奶的量大约减少一半；这个过程大概要 75 分钟。在煮牛奶的同时，粗略地切一下开心果。

牛奶减少一半或以上，出锅，把豆蔻荚挑出来扔掉。把米布丁倒入碗内。加入糖品尝一下，如果你想让它更甜的话，可以再多加点糖。把刚刚切好的开心果放进去，搅拌均匀，然后自然冷却。

用保鲜膜把碗包好，放入冰箱冷藏四个小时以上，或者过夜。

要上桌之前，用勺子把米布丁分到一人份大小的碗中。如果需要，用金银箔等装饰一下。

如果喜欢的话，也可以再撒上一点开心果碎。

糯米布丁

（适用于乳糖不耐受的食客）

675 毫升（3 杯）香草味米露

200 克（1 杯）糯米

150 克（3/4 杯）砂糖

1 茶匙新鲜生姜，磨成姜蓉

1 块柠檬皮

1/4 茶匙盐

1个香草荚，切开刮出香草泥（留着豆荚，放入糖

罐或者伏特加瓶里面保存）

2枚大个鸡蛋

1颗蛋黄

1茶匙切成碎末的糖渍柠檬皮

2茶匙黑朗姆酒

将两杯米露、糯米、半杯糖和姜蓉、柠檬皮混合，倒入一个中等大小的深煮锅里。将米露烧开。快速转小火，加盖焖制。煮大概20~25分钟的样子，等大部分米露都被米粒吸收，就可以关火了。开盖，晾差不多30分钟。

在一个中等大小的碗里，将剩下的米露、糖、盐、香草泥、全蛋和蛋黄搅打均匀。混合均匀后过筛，倒入深煮锅。中火加热，一直搅拌到液体可以挂在勺子的背面，大概8分钟左右。出锅。

在液体中加入糖渍柠檬皮、朗姆酒和凉凉的米饭，搅拌到混合均匀为止。混合物应该是比较稀的。

把布丁倒入一个盛菜用的碗里，或者分成六小份倒进刷过油（最好是杏仁油或者高品质的橄榄油）的蛋糕杯里，确保容器翻过来的时候做好的布丁能顺利滑出来。脱模前要冷藏几个小时。在室温下回温一个小时后，布丁就可以端上桌了。

柔软的椰子或者芒果雪葩是很好的搭配，最好在上面洒上几滴浓缩陈年香醋。

6人份。

炒 饭

炒米饭一般是使用剩饭做的。炒饭应该是最常见的处理剩饭的方法，不仅方便快捷，而且其他剩余的食材也可以放进米饭里一起用掉。剩饭有独特的特质，也自然有相应的制作技巧来突出这些特点。所有的食米文化都会有自己处理剩饭的方法，而这些菜肴又成了其烹饪传统的重要部分。

印尼炒饭

这个菜谱是根据中式炒饭改良的印尼和马来版本。我的配方改良自迈克尔·弗里曼的《大米之地：东南亚美食王国》（伦敦，2008）。

3 茶匙菜油

3 瓣大蒜，切碎

4 颗红洋葱，切末

150 克（约 5 盎司）生虾仁，去壳

150 克（约 5 盎司）鸡肉，切成 5 厘米（2 英寸[①]）

大小的肉丁

1~2 茶匙生抽酱油

400 克（约 14 盎司）隔夜冷饭

---

① 英寸为长度单位，1 英寸约为 2.5 厘米。

4枚蛋

2棵小葱，保留葱绿切末

3个中等大小的新鲜辣椒，去籽切碎

1茶匙欧芹，切碎

3棵香菜，去叶切碎

一小撮盐

一小撮胡椒碎

铁锅内烧油至略微冒烟，调成中火，下蒜末炒至金黄色。加入红洋葱，翻炒至褐色。放入虾仁、鸡肉和酱油，翻炒至虾仁变红、鸡肉变白。下米饭，和虾仁、鸡肉炒匀，不停炒动几分钟，等米饭变热。盖盖子，放在锅中备用。另起一锅，倒油煎鸡蛋，注意不要搅动蛋黄。盛出备用。

将米饭盛到一人份的盘中，在上面撒上小葱、辣椒、欧芹和香菜，用盐和胡椒调味，最后在每份饭上面放上煎鸡蛋。

菜豆烩饭

菜豆烩饭是一种起源于非洲的用大米和豆类/木豆制作而成的美食，有着几百甚至上千年的历史，但下面的菜谱是一个现代的版本，用的也是现代大米。非洲和亚洲的大米混合着不同的豆类就能做出独具国家特色的美食，成为很多国家烹饪历史的一部分。像卡伦·海斯、杰西卡·哈里斯和詹姆斯·麦克威廉姆斯这样的非洲或者非裔美国饮食文化的学者都曾经推测过这道菜名字的由米。然而，现在还没有统一的结论。"Arroz con frijoles"是菜豆烩饭的拉丁名称，在加勒比、墨西哥和南美洲到处都能看到。

这个菜谱改良自杰西卡·哈里斯的《宾至如归：非裔美国人的烹饪传统》（纽约，1995）

1 磅（450 克）干的黑眼豆（豇豆）

1/2 磅（225 克）咸猪肉

1 夸脱（950 毫升）水

1 枝新鲜百里香

盐和新鲜研磨的黑胡椒，用来调味

1.5 杯（150 克）生的长粒大米

3 杯（675 毫升）热水

择净黑眼豆，去掉泥土和石子。用水浸泡至少四个小时，也可以隔夜。在一口结实的大砂锅里炒制咸猪肉，煸出肥油。等咸猪肉变脆之后，加入黑眼豆和一夸脱的水、百里香、盐和黑胡椒，加盖用文火慢炖 40 分钟左右。再调一下味道，继续炖到豆子软烂。加入大米和足以没过食材的三杯热水，小火焖煮至汤汁全部被吸收、米粒松软。趁热装盘。

在新年那天，一些家庭会在菜豆烩饭中放一枚硬币，吃到的人就会有一整年的好运。

牡蛎饭：一份古拉菜谱

这是一份来自低地国家的食谱：美国佐治亚州海岸外的那些沿岸岛屿住着古拉族的非裔美国人，古拉正是"食米之人"的意思。食谱来自萨莉·安·罗宾逊和格里高利·雷恩·史密斯所著的《古拉家庭烹饪：都福斯科之道》（2003）。

4 条培根

1 汤匙食用油

1 个大洋葱，切碎

1 个中等大小的绿色灯笼椒，切碎

2 汤匙面粉

3 杯（675 毫升）热水

盐和黑胡椒用于调味

2 杯（400 克）生米

1 夸脱（1 千克）去壳牡蛎，沥干水分

在一口中等大小的锅里将培根煎脆。把培根夹出来，锅中剩余肉脂肪油。再加入一点食用油，洋葱和青椒炒至洋葱变得透明。把蔬菜盛出，留下食用油和肉脂肪油。用底油把面粉炒成焦褐色，把炒好的培根、洋葱和青椒倒回锅中。加水，用盐和黑胡椒调味，等到水沸后转小火焖煮15 分钟，时不时搅拌一下，让锅中形成薄薄的肉汁。将大米冲洗沥水几次，冲洗牡蛎，将两种食材放入锅中。充分混合后加盖慢炖 30~45 分钟，时常搅拌一下。搭配一些蔬菜小菜，就是一顿饭了。

米华夫饼

改良自范妮·梅里特·法默的《波士顿烹饪学校

厨艺指南》（1896）

1.75 杯（600 克）面粉

4 茶匙发酵粉

2/3 杯（90 克）隔夜冷饭

1/4 茶匙盐

1.5 杯（340 毫升）牛奶

1 汤匙融化黄油

2 汤匙糖

1 枚鸡蛋

混合并过筛干性原料；放入凉米饭并搅拌；加入牛奶、打发的蛋黄、黄油和硬性打发的蛋白。接下来和华夫饼的制作方法一样。

烤肉饭

以下菜谱改编自 1971 年得克萨斯州休斯敦美国水稻协会出版的小册子。取悦男性的食谱会以"没有哪个男人每晚会喜欢一样的东西！"的字样开头。米饭食谱包括早、中、晚三餐。这个指南的主要目的就是给读者的家常菜里添加一些新鲜味道。"餐盘中的蔬菜或精致晚宴的食谱。很好吃 —— 试试看吧!"

1 杯(125 克)切碎的小葱,保留葱绿

1/2 杯(60 克)青椒碎

2 汤匙黄油或者人造黄油

3 杯(400 克)用牛肉汤现煮的米饭

3 汤匙切碎的甜椒

盐和胡椒

在黄油中烹制小葱和青椒,直到两种食材变脆。加入米饭和甜椒,轻轻颠一下锅。根据口味调整味道,配上你最爱的烤肉即可食用。

6 人份。

蒸米饭 II

改良自格洛丽亚·布莱·米勒的《一千道中国美食菜谱》(纽约,1970)

洗净大米。放入锅中,加入足量的水。煮 5 分钟,其间搅拌几次。沥干水分。

将米码放在盖着屉布的竹制蒸笼上。用筷子或者叉子在米饭上捅一些小孔,让水蒸气能散出来。

盖上盖子,中火蒸 20 分钟。

米勒还写到,第二步中沥出的水可以加点糖,当作稀

米汤来喝。

罗斯玛塔米[①]

改良自杰弗里·奥尔福德和内奥米·杜吉德的

《大米的诱惑》(纽约,2003)

这种来自印度南部特殊的煮成半熟的红米只进行了部分碾磨,部分红色的米麸层没有被打磨掉。虽然预煮过,但还是要洗掉第一次烹饪时形成的碎屑和浮渣。这种米有一种鲜味或肉香,米粒能保持粒粒分明。

2杯(400克)米

3杯(675毫升)水(如果是用电饭煲,就用

2.25杯,也就是500毫升水)

在流水下将米彻底冲洗干净,水开始会是棕红色的。放在筛子里沥干水,挑拣一下米,将硬实粒或者不规则的米粒挑出去。把米放进一个结实的中等大小的锅或者电饭煲里,加水。如果是用锅的话,就先大火煮沸,快速搅拌一下,然后敞着口煮3~4分钟。再搅拌一下,然后盖上锅盖,把火转为中小火。焖5分钟,调成文火,盖盖再煮12~15分钟。关火,不要掀开盖子,用蒸汽焖10~15分钟,用一把木铲稍微搅拌一下。做好的米饭应该是紧实,甚至有点

---

① 罗斯玛塔米生长在印度南部喀拉拉邦帕拉克拉德地区,以其粗糙度和益于健康而闻名。

Q 弹的，煮透了的。

如果是用电饭煲的话，打开电源，盖上盖子煮就好了。当它的电源自动跳转以后，放置其继续焖10~15分钟再开盖翻动米饭。

鸡肉粥

许多人的早餐，适合所有人的食物——粥，经常会就咸菜、酱油、咸花生米、萝卜干、腌姜、梅菜、香肠、咸鱼和任何手边的剩菜。这道菜谱改编自林相如和廖翠凤编写的《中国美食》（纽约，1977）。

100 克（1/2 杯）稻田米（短粒或者其他淀粉
含量高的大米）
6 杯（1.35 升）鸡汤
1 块鸡胸肉
1/2 平勺盐
2 汤匙清水

把米洗净，放入鸡汤中煮沸。转成小火，慢炖两小时左右。在煮粥的时候把鸡胸去皮剔骨，顺着纹理切片。用切肉刀背把鸡肉片拍平一些。加入盐和清水。当稀饭准备上桌时，关火。把鸡肉片搅拌均匀，放进粥里，放上 3~4分钟。盛到碗里即可。

马来西亚黑糯米粥

改良自夏麦尼·所罗门的《亚洲美食百科全书》

（波士顿，1998）

220 克（1 杯）黑糯米

1.5 升（3 品脱）水

60 克（2 盎司）棕榈糖

2 汤匙砂糖

2 片班兰叶

6 颗桂圆干

250 毫升（8 液量盎司[①]）椰浆奶油

1/4 茶匙盐

换几次水把米洗干净后沥干水分。把米和量好分量的水放入一口结实的煮锅中，煮至沸腾。盖盖子焖煮 30~40 分钟，时常搅拌一下避免米粘在锅底。加入棕榈糖、砂糖、班兰叶和桂圆干（如果桂圆是带壳的，要把壳剥下来扔掉）。如果粥太稠了，可以加一点热水。一直煮到米粒变得非常柔软。与加了盐的椰浆奶油一起，趁热食用。

---

① 液量盎司是一种容量计量单位，常用的有英制和美制液量盎司两种。1 英制液量盎司约 28.41 毫升，1 美制液量盎司约 29.57 毫升。

糙米欧洽塔

改良自 www.massaorganics.com

1/2 杯（100 克）糖

1 袋（7 盎司或是 350 克）无糖椰子片（如果用的

是加糖的椰子片的话，就减少糖的用量）

3/4 杯（150 克）糙米，用水浸泡一夜之后沥干水分

1 杯（135 克）烘焙过的去皮杏仁

1 条肉桂棒

1/4 杯（55 毫升）香草米露

将糖和 5 汤匙水倒进一口小号平底锅内，加盖用中火煮 4~5 分钟，不时晃动锅子确保糖完全溶化。把糖浆倒进一个小碗里，凉凉。

将椰子片和一杯半的水加入搅拌机，搅打至顺滑。将液体用一个细筛过滤进一个碗里，用橡胶刮刀按压筛网上的固体，尽可能多萃取出一些椰奶。放在一旁备用。

将糙米、杏仁、肉桂和两杯清水加入洗干净的搅拌机里，搅打至顺滑。用铺有纱布的筛网把混合液体过滤到一个碗中，按压一下固体，尽可能多地提取液体，再把过滤完的固体倒回清洗干净的搅拌机中。再将 3/4 杯的椰奶、糖浆、米露和两杯冰块放入搅拌机，将冰块打成冰沙，饮料打到起泡。分到 2~4 个玻璃杯中，尽快饮用。

2~4 人份。

粢毛肉圆

改良自林相如和廖翠凤的《中国美食》（纽约，1977）

2.5 盎司（6 平勺）糯米

1 平勺盐

1/4 磅（100 克）肥肉末

1/4 磅（100 克）瘦肉末

1.5 茶匙料酒

0.5 茶匙糖

2 茶匙生抽

1 平勺玉米淀粉

0.5 平勺味精

油

5 茶匙酱油

3 茶匙醋

将糯米放入一个 2 品脱大小的量杯里（或者一个 4 杯大小的耐热玻璃碗），加水到 24 液量盎司（3 杯）的刻度线。泡大概 45 分钟，然后沥干水分。将糯米和盐搅拌均匀。

在另一个碗里，把肥瘦肉末、料酒、糖、酱油、玉米淀粉、味精和 1 汤匙水搅拌均匀。把调味料抓进肉馅里，然后团成直径 1 英寸大小的肉丸。用糯米把丸子包起来，然后摆在一个涂了油的盘子上。盖紧盖子蒸 30 分钟。在一个小碟子里放上酱油和醋，和丸子一起端上桌。

甜米糕

改良自雷纳尔多·亚历杭德罗的《菲律宾烹饪书》

（纽约，1985）

这本菲律宾菜谱还配了一条提示，说菲律宾的用餐礼仪包括把所有餐具或菜品一起端上桌，不论甜咸。因此，这道菜不完全是一道甜品。

4 杯（560 克）甜米粉

1.5 杯（150 克）糖

2 罐椰奶

半茶匙盐

将甜米粉放在烤盘上烤一下。将糖、椰奶和盐拿去煮沸。加入 3 杯烘烤过的甜米粉。好好混合一下，不停搅拌，直至其变得很稠。出锅，将米团倒在一个铺满剩余甜米粉的案板上。用擀面杖将米团擀成四分之一英寸薄厚的饼，切成方块状。再将剩余的米粉擀进面饼里。

可制成 15~20 块小蛋糕。

米 粉

米粉可以是湿粉或者干粉，可长可短，可粗可细，甚至可以是做好的米皮，任你切成不同的形状。中国和泰国是最大的米粉出口国。有些是用大米粉做的，有的是用糯米粉做的，还有其他的各种粉类（像是木薯或者玉米淀粉），创造出截然不同的弹性和质地。

鲜米皮可以切条或者包东西吃，但一定要在变干前尽快食用。干粉（也称作粉丝）要在常温的水里泡 10~20 分钟，变软后再烹制，不然就会变成糊状。如果是要做汤的话，为了去除多余的淀粉保持汤的干净，米粉可以在泡软后在锅里焯一下。

新加坡米粉

改良自科琳娜·翠音的《一日一面：从拉面到米粉的美味亚洲食谱》（旧金山，2009）

这本菜谱在美国的中餐馆里很受欢迎，经常会用到剩的广式叉烧肉和咖喱粉，后者也反映了印度给新加坡带来的影响。

8 盎司（225 克）干的细米粉，在水中泡软

24 只小虎虾，去头剥壳，去虾线

3 汤匙植物油

1 个小洋葱，切成小丁

0.5 杯（75 克）现剥豌豆或者解冻的冷冻豌豆

2 汤匙印度咖喱粉

6 盎司（175 克）广式叉烧肉

1.5 汤匙鱼露

海盐和现磨黑胡椒

6 片香菜叶，切碎

在一锅沸水中将米粉煮软，大概 10 秒钟就好。把米粉盛进一个碗里。在煮面水里煮虾，一分钟左右。

在平底锅或者铁锅上烧热一汤匙油，将洋葱丁炒至金黄，大概 3~4 分钟。再加入 2 勺油、米粉和豌豆。将咖喱粉撒在上面。搅拌均匀让米粉覆盖上黄色。加入叉烧肉、虾、鱼露，翻炒 5 分钟左右。用盐和胡椒调味，出锅后撒上香菜。

6 人份。

致

谢

在此特别鸣谢在我撰写本书期间给我提供莫大帮助的人们：杰伊·巴克斯代尔、菲尔·布鲁诺、罗伯特·卡马克、艾米·科尔、多莉·埃里奇、巴里·埃斯塔布鲁克、苏珊·法斯、亚历克斯·加西亚、琼·吉尔达内拉、詹妮·休斯顿、J. J. 雅各布森、雷切尔·劳丹、莫妮克·利农、夏洛特·林德伯格、麦玲、扬·隆根、凡妮莎·卢辛、丹妮尔·玛顿、加里·玛顿、西蒙·玛顿、乔安娜·麦克纳马拉、洪阮、玛格丽特·哈佩尔·佩里、莫里森·波尔金霍恩、朱迪·鲁西诺洛、玛丽·西蒙斯、安迪·史密斯、简·斯坦尼基、里克·斯坦、加里·陶贝斯、劳拉·韦斯、大卫·韦克斯勒、斯塔西娅·威尔基、莎拉·沃默，特别是埃德·史密斯。

我还要谢谢纽约公共图书馆、密歇根安娜堡的柯莱蒙兹图书馆和其他各个城市的图书馆。没有这些地方，我们的精神生活定将变得无比匮乏。网上的数据库虽然也很重要，但总归是不够用的。

承蒙很多人的帮助，书中的任何不足，定是我的过失。

Rice

A Global History

Renee Marton

Contents

Introduction

Without rice, even the cleverest housewife cannot cook.

Chinese proverb

It is possible to consume rice on a daily basis wherever you are in the world. In fact, two-thirds of the world's population already does this, chiefly in countries in which rice agriculture has long been established. Rice is also increasingly consumed in countries where large numbers of immigrants from rice-based countries have settled. China and southern Asia, northern India and other Asian countries from Indonesia to Myanmar to Japan, and western and central African countries: these areas are the birthplace of rice farming. In other places, rice arrived as an immigrant grain. Early on, the profit motive and the need to feed labourers (not necessarily simultaneously) were the twin drivers of rice cultivation and commerce. Human migration, whether voluntary or forced, paralleled rice migration. Both rice and people adapted to their new locations.

If you travel from New York to Guangzhou, breakfast will likely be *congee* or *juk*. This rice porridge, eaten by millions daily, is typically made from rice left over from the night before. You might also enjoy *congee* in Sacramento, California, where the descendants of Cantonese immigrants, who travelled to the U.S. because of the California Gold Rush of the 1850s, have remained. Rice was imported as sustenance for the 40,000 Chinese labourers in California. As an industry, however, rice agriculture in California did not begin in earnest until the late nineteenth century, with commercial rice farming accelerating in the 1920s. In 1850, when California became a state, most rice was imported from China. However, in 1950, rice agriculture in the Sacramento Valley was well established. And by 2008, 50 per cent of California-grown rice was being exported to Japan, Korea, Uzbekistan and Turkey.

It was Cantonese immigrants who opened the first Chinese restaurants in the U.S., which catered primarily to Chinese clientele. Americans slowly developed an interest in 'oriental' foods, and some Chinese cooks began working in private homes. For these Cantonese Chinese, unadorned rice accompanied almost all their meals (rice is usually served plain in countries where it is the staple starch). What we call 'fried rice' arose through the judicious use of leftovers. Today, fried rice can be ordered from menus as a dish on its own, representing an adaptation by and for non-Asians. After immigration laws were relaxed in the U.S. in 1965, Chinese diaspora immigrants from Taiwan, Hong Kong and Fujian expanded the definition of 'Chinese' in New York, San Francisco, Los Angeles and other cities. Their rice dishes came with them.

If you celebrate New Year's Day in Charleston, South Carolina, hoppin' John will likely be a part of the meal. This traditional African meal of rice and cowpeas, or black-eyed peas/beans, comes from the culinary repertoire of West Africans. It was brought to the British, Dutch, French, Spanish, Portuguese and the as yet unnamed USA (where rice became a premier crop) colonies by slaves, as well as to some Caribbean islands, Brazil, Peru, Cuba and Mexico by the workers in sugar cane, cotton, tobacco, indigo and other colonial plantations. Indians also came as indentured labourers to the West Indies, followed by the Chinese and others. Rice, imported at first for slaves and labourers, became a commercial enterprise. If you eat hoppin' John today, you are very likely to be of African or Caribbean descent, or both. And while you could be located in the southeastern U.S., the Caribbean islands or Mexico, you might equally be based in Detroit, Michigan, or Gary, Indiana.

You meet a friend for sushi and sake. Where are you? Tokyo? Perhaps, but wait: you hear Portuguese and English—it turns out you are in São Paulo or London. Sushi is almost global in its urban reach, although it is a relatively late developer in the world of global rice-based foods. And consider the California roll. This inside-out roll, sometimes made with brown rice, includes avocado, cucumber, carrots, omelette and herbs, but no raw fish. California rolls are found on the menus of elegant Japanese restaurants in Singapore and Kosher restaurants in Shanghai. In culinary schools in

Tokyo chefs are even taught how to prepare them 'properly'.

While on the plane back to New York you read about famous rice-based meals in your in-flight magazine: paella, risotto, biryani and pilaf. These four rice meals are prepared at homes and in restaurants, at outdoor markets, on food trucks and at festivals; they are linked to Spain, Italy, India and Iran respectively. Recent historical research indicates that the origins of these dishes go as far back as the Moghul dynasties and Islamic traders.

Once back in New York, the Bangladeshi taxi driver at the airport offers you a snack of *jhal-muri*. This Kolkata speciality consists of puffed rice (*muri*) seasoned with lemon and coriander and mixed with peanuts, chopped onions and chilli. On one side of the New York street, South Indian street vendors sell *dosas*, crêpes stuffed with rice and lentils, on the other Pakistanis sell chicken rice, curry rice and lamb biryani.

While the colonial relationship between Britain and India ended in 1947, the culinary ties between these two countries had been evolving for more than a century, as immigrants from India came to the UK before, during and after independence. Many Brits returning from India longed for the flavours of that country's cuisine, and a few were able to satiate this appetite courtesy of the Indian wife or cook they brought back into the country with them. Indian immigrants integrated their culinary backgrounds into their new lives, and into ours. Street vendors and restaurants, and later prepared food manufacturers and supermarkets, accelerated this transmission.

You feel ill from overeating. To settle your stomach, you prepare a soothing bowl of cream of rice (remarkably similar to *congee*). You also drink *horchata*, a cooling rice-based drink from Central America and Mexico. You start a diet of puffed brown rice crackers and rice-based green tea, with an occasional Budweiser beer; yes, rice is a principal ingredient of Budweiser. You make Rice Krispie treats for your children, and feed rice to your dog. Is rice in pet food? You guessed right. Rice is everywhere, and these examples are just the tip of the 'riceberg'. Finally, when you feel recovered, you prepare a rice pudding, studded with raisins and pistachios, to take to a dinner party for dessert.

Chapter 1
Country and Culture

Whether steamed in banana leaves, simmered in pots or cooked in an electric rice cooker, rice is frequently a white canvas on which culinary cultures are painted. *Kimchi*, soy sauce, salt pork, dried fish, yams or oysters: these seasonings and/or accompaniments to rice imply heritage origins. In rice cultures, rice is testimony that you are having a meal, and is the main source of caloric energy in the diet. White, polished grains are generally the preferred form of rice but partially milled rice is also widely consumed: it is less labour-intensive to produce, costs less to procure and is more nutritious than white rice. When ground, rice can form the basis of noodles, flatbreads, cakes and crackers. Rice flour is used as a thickener for sauces, puddings, sausages, baby food and pet food, and is also used in cosmetics. Other products in which rice is a constituent include rice paper wrappers, vinegar, miso, bran oil, beer, rice 'milk' and 'wine'. Puffed, popped and flaked rice is used to make cereals, snacks, crackers, pastries and breads. Stalks and husks are woven into sandals and mats, or are burned as fuel.

How Rice Grows

Rice is a highly adaptable cereal grass that grows in most environments, but not necessarily prolifically. Irrigated rice accounts for 50 per cent of cultivated rice, and represents 75 per cent of the almost 700 million tons of rice harvested in 2010 (of which about 30 to 35 per cent is lost to husking, milling and polishing). Wetland rice begins as nursery seedlings that are later transplanted to paddies. This is an arduous and physically demanding process that produces one, two or three crops per year (depending on rice type, location and climate). Wet rice agriculture is found primarily in tropical latitudes: southern China, Southeast Asia, India and other Asian countries or regions; North America, mainly the U. S. and Mexico, as well as countries or regions in Africa and South America. Rain-fed lowland rice accounts for another 20 per cent and upland rice for 5 per cent. Upland rice agriculture, also called dry cultivation, is found chiefly in South America and Africa. Finally, there is deep-water rice, which can grow in water depths of 50 cm or more, and is a significant crop in Bangladesh and other areas with deeply flooded river valleys. Rice plants grow rapidly above rising water, up to 4 m.

As the world's population increases and cities expand, rice producers strive to keep pace with demand. Along with higher yields and soil-friendly irrigation methods, vertical agriculture may provide a partial solution: essentially a variation on land-based terraced rice paddies, this approach involves growing rice in vertical layers up the sides of buildings or in giant greenhouses. Newer high-yield rice strains are being developed to accommodate increasing demand. 'Organic' rices are being explored for flavour and sustainability, while the latest news about arsenic being found in rice is being evaluated and methods for reducing it explored. While arsenic is a naturally occurring mineral that is absorbed through plant roots, the levels have been increasing recently, especially in California rice. This is being investigated at present, but so far there are no definitive causal explanations.

The Cultural and Culinary Value of Rice

White rice lends itself to multiple preparations and cultural heritages. If you ate a *puso* (rice cooked in a leaf packet), before 1965, you were probably Filipino and in Manila. After the relaxation of immigration quotas in the U.S. in 1965, you might very well have been in Daly City, California, where large numbers of immigrant Filipinos settled.

Brown rice, a complex carbohydrate with little fat, chollesterol or sodium, contains eight amino acids, B vitamins, iron, calcium and fibre. It is very nutritious and has a chewy texture and nutty flavour. So why do most people prefer white rice? White rice stores well once the bran and germ are removed, cooks quickly and is inexpensive, digestible and satisfying. The colour white also has a long-standing symbolic value: purity, cleanliness, status and quality.

Cooked rice can be very soft, as in *congee*, or as hard as

the *tahdig* (golden crust) at the bottom of a pilaf. Rice can be salty or sweet, soft or crunchy, hot or cold, and even used as a tool itself: sticky rice is especially amenable to being pressed by one's fingers into a mini bowl, which is then dragged through sauces and picks up small pieces of meat, fish or vegetables. It is endlessly versatile.

In areas with the highest rice consumption rice provides two-thirds of the daily calorie intake of the people who live there. Four more traits explain why rice has such loyal fans: first, it complements other flavours, as in pilaf or rice pudding, while maintaining its own flavour and texture. Second, rice is a buffer for spiciness, as in a vindaloo, curry or gumbo. Third, rice retains moisture, enhancing succulence, as in stuffed grape leaves or boudin sausage. And finally, the inclusion of rice generally reduces meal costs.

Foremost a staple food for people living at or near subsistence levels, rice has recently become an upscale item in 'developed' countries. Statistics make clear that as people become more affluent they consume more animal protein. As discretionary income rises, high-status protein foods are sought out. Yet in industrialized countries, health concerns currently trump status displays and brown rice intake is increasing, with concomitant decreases in meat and fish. However, as the population increases, so will the need for rice, and that rice will probably be white.

According to the International Rice Research Institute (IRRI) in the Philippines, individual rice consumption by weight was highest in the following twenty countries in 2010.

Rice is measured in kilograms consumed, per person, per year. China, India and Indonesia are not at the top of this list in terms of consumption per person, but are still the leading producers and consumers of rice in the world because of the size of their populations.

Rice consumed yearly per capita in kilograms

Brunei	245
Vietnam	166
Laos	163
Bangladesh	160
Myanmar	157
Cambodia	152
Philippines	129
Indonesia	125
Thailand	103
Madagascar	102
Sri Lanka	97
Guinea	95
Sierra Leone	92
Guinea-Bissau	85
Guyana	81
Nepal	78
Korea, DPR	77
China	77
Malaysia	77
Korea, Republic of	76

In comparison, annual rice intake in Brazil in 2008 was 44 kg per person (partially thanks to sushi consumption, particularly in São Paulo), and in the U.S., 11 kg per person.

Because Asian rice (*Oryza sativa*) is the dominant rice family from which most cultivars are descended, and as cultivars vary greatly in sensory attributes, *terroir*—the particular environment in which a plant is grown, including climate, geology and geography—is important not only for plant genetics, but for flavour, colour and aroma. There are more than 115,000 rice cultivars. Since rice agriculture is very labour-intensive and also requires vast quantities of water for maximum yields, you might wonder why people for whom rice is the daily staple would go to so much trouble to produce it. In this question lies the answer: people are loyal to the grain precisely because they have worked so hard to grow it. (This preference continues when people move to cities, but is expressed in different ways: for example, having a meal without rice in a city still means that a 'true' meal has not taken place, but perhaps a snack). And each migrational group wants its 'own' rice to be available. The fieldwork is worth it: rice yields and calories consumed are higher per acre than those for wheat, maize, soya beans and millets.

Types and Forms of Rice

Which rice do you prefer: brown, Basmati, sticky or Uncle Ben's? While we are familiar with these descriptions, they are not a real taxonomy, since rice varies in grain size, shape, colour, stickiness, taste and aroma. Remove the hull (or husk, the protective shell that covers the grains) and you are left with brown rice, which still retains the bran, germ and nutrients. Removing these elements gives you white rice, which is almost pure starch.

Basmati is a long-grain, usually aged, usually white, aromatic Indian, Pakistani or Bangladeshi rice, highly valued for its popcorn-like aroma and ability to fluff well after cooking. The grains remain separate from one another.

Sticky rice grains can be long or short and cling together after cooking. Sticky rice may be used as a 'tool' to literally 'pick up' other parts of the meal. Some say sticky rice is more filling than other types of rice. Long-grain sticky rice is found in Thai, Laotian and southwestern Chinese communities.

Uncle Ben's is the brand name of 'converted' rice, in which nutrients from the bran covering are 'pushed' into the core during parboiling, retaining 80 per cent of the nutritional benefits. Harvested and packaged in the U.S. by Masterfoods, Inc., a subsidiary of Mars Inc., Uncle Ben's Rice is a global brand.

Indica rice grows in tropical and subtropical regions and accounts for more than 75 per cent of global trade. Indica rice cooks dry, and the grains remain separate.

Japonica rice typically grows in cooler climates and accounts for about 10 per cent of the global rice trade. Japonica grains stick together somewhat and can be eaten easily with chopsticks.

Aromatic rice, primarily Jasmine from Thailand and Basmati from India, Pakistan or Bangladesh, accounts for more than 10 per cent of global trade and typically sells at a premium in world markets. All are long-grain, with distinctive aromas, and fluff well after cooking.

Glutinous rice from Southeast Asia is used in desserts and ceremonial dishes and accounts for 5 per cent of global trade. Glutinous rice is moulded into logs and balls to be used for dipping into sauces and curries or is made into 'wrappers' for sweet or savoury fillings. There is no connection between glutinous rice and gluten, the protein found in wheat, rye and barley.

As with most attempts to classify such a diverse and adaptable cereal grass, the four categories of rice the USDA recognizes—indica, japonica, aromatic and glutinous—exclude javanica rice, which falls somewhere between the indica and japonica varieties. The USDA also omits the fact that in northern Thailand, in Laos and in Yunnan province, China, glutinous rice is not only used for sweets but is also eaten as part of the main meal. American versions of Jasmine and Basmati rice are called, among other names, Jazzmen and

Texmati: while they may be similar to Jasmine and Basmati, they are not included in the classification system, although that may change through lobbying in Washington, DC.

Rice has two main starches: amylose and amylopectin. The ratio of one to the other determines the fluffiness or stickiness of cooked rice. Sticky rice has a rounder, shorter grain and very limited amounts of amylose. Once cooked, sticky rice can be moulded into a slab, or rolled into balls for dipping. Rice balls are used in the same way that a slice of baguette or wedge of naan would be in wheat-based cuisines: for picking up sauce or pieces of meat or fish. Sticky rice can be used as a 'wrapper' for vegetables, bean paste or fruit and for dumplings.

Glutinous rice is a staple food in southern China, Laos and Thailand, where rice is soaked and steamed. In Japan, Korea and northern China, somewhat less sticky rice is preferred, while medium sticky rice is most desired in Indonesia, the Philippines, Malaysia and Vietnam. These rices are of the japonica or javanica variety and are in the mid-range of the sticky to fluffy rice continuum.

While Thailand, Vietnam and the U.S. are the top three exporters of indica rice, sticky rice fans are increasing worldwide. Sweet or *mochi* rice is another form of glutinous rice, which is favoured and grown in southwest China, Southeast Asia and Japan and used for puddings and as coverings for desserts. Glutinous and non-glutinous rice is sometimes mixed together in order to achieve specific tastes, textures and levels of malleability. When eaten plain, glutinous rice is sometimes served in the woven steamer basket in which it was cooked.

Short-grain rice is traditionally used in paella, but medium rice is often used as well. A paella pan is wide and shallow with two handles. A mixture of aromatics and/or meat is sautéed in the pan, dry rice is stirred in, and hot liquid is added. A classic version would contain rabbit or snails; a modern version might include shellfish, chicken or vegetables. Paella is traditionally cooked on an outdoor fire or grill and should remain uncovered during cooking, until all the liquid is absorbed. Once the liquid has been absorbed, the heat is increased and a *soccarat* forms: a layer of browned crunchy rice at the bottom of the pan. A contemporary twist involves sautéeing the aromatics and rice first; then the other ingredients and liquid are added to the pan, and it is covered and placed in the oven to finish cooking.

At the other end of the rice continuum are long-grain rices: the so-called basmatis and jasmines of the world. (True Basmati and Jasmine rices, with a capital first letter, come from India, Pakistan and Bangladesh (for Basmati) and Thailand (Jasmine) respectively—specific land areas that produce specific rice with protected names. When long-grain rice not of this exact type but of a similar style is referred to, however, the lower case is used: basmati and jasmine.) After cooking they elongate to two or three times their raw length. The grains remain separate after cooking. Soaked, rinsed until clear, boiled like pasta, or simmered or steamed over a very low heat in a covered pot or pan: these are the rice-cooking methods for preparing this rice plain. Dishes such pilafs and biryanis, soups and baked rice dishes, where other ingredients are added to the rice before serving, employ different cooking methods. In pilafs, for example, the rice is first sautéed in fat (originally fat from fat-tailed sheep, a prized ingredient in Persia) before being combined with a hot liquid such as stock. Other ingredients are added to the rice at different stages of the cooking process. A pilaf with chicken, raisins and chickpeas will end up with mixing all the ingredients together simultaneously. Pouring ghee (clarified butter) down a hole in the centre of the rice, sealing the pot with a flour/water paste or a towel and letting it steam over a low heat creates a *tahdig*: a crunchy brown crust on the bottom of the pot, which is very similar to the *soccarat* in paella. Alternating layers of lamb and rice, seasoned and cooked separately and almost cooked through before they are combined in layers, gives us biryani rice dishes, of which there are multiple variations. In China, the crunchy rice crust is called *guo ba*, in Korea *nurungji* and in Senegal *xoon*.

Developments in Rice

Rice production was a beneficiary of the Green Revolution of the 1940s to 1970s, the scientific developments in agriculture that improved crop yields and disease resistance and are thought to have saved more than 1 billion people from starvation. In 1970, the Nobel Peace Prize was awarded to Norman Borlaug—the first for an agronomist—for his pioneering work in developing high-yield and semi-dwarf varieties of wheat and maize in Mexico. His methods were used later for growing rice and wheat in Asia, through the auspices of the International Rice Research Institute (IRRI), which has a repository of thousands of rice species, as well as funding ongoing research projects to increase the yield of rice plants while reducing pests and plant disease outbreaks. The IRRI developed varieties of high-yielding rice seeds. This work continues today with research into rice with higher yields and lower water requirements. Korea became self-sufficient in rice production after switching to high-yield rice varieties and modern management methods. India has begun to use laser levelling—using lasers to flatten the land and create rows—for rice fields. This is less arduous than physically levelling the land, and the resultant fields use less water during irrigation.

Rice paddy irrigation results in the production of methane gas, a contributor—albeit a small one—to global warming. Land used for biofuel production (instead of harvesting plants for human consumption) and genetically engineered (GE) rice: these controversial topics remain tied to the future of rice. Some countries have returned to traditional rice agriculture, even as they are using modern seeds (Indonesia, since the 1970s). Despite record harvests in the last ten years, coastal areas, many of which are in rice-producing areas, have already been inundated as a result of climate change.

Protecting, preserving and planting older breeds of rice is one way of avoiding future monocultures that might be susceptible to particular environmental problems or disease. The IRRI maintains a gene bank of rice cultivars as an attempt to safeguard against the practice of monocultures.

Other scientific developments in rice may solve other potential problems. The Central Rice Research Institute in Cuttack, India, in 2010 developed a rice variety with very low amylose: aghonibora. The advantage of lower starch levels is that less water is needed to produce ready-to-eat rice. It has only to be soaked in warm water for 30 minutes before it is ready to consume. Floodplain rice is also being looked into as an alternative grain for the future, since climate change may lead to flooding in many parts of the world. The Cuu Long Delta Rice Research Institute in Vietnam was established to develop high-yield rice strains with a short planting season and the ability to be planted away from the Mekong Delta floodplain, which is expected to disappear as sea levels rise. In the U.S., the Rice Research and Extension Center Institute in Stuttgart, Arkansas (Arkansas produces 50 per cent of U.S. rice), maintains efforts to keep American rice competitive in the global market, and to research newer varieties for the future.

Through the research and development of sustainable ways of farming, rice will remain available to those who already depend on it for daily sustenance, as well as for the population of the future.

Chapter 2
The Old World

Approximately 15,000 years ago, the ability to domesticate plants turned some hunter-gatherers into agrarian communities. Certain cereals were preferred: wheat, barley, millets, sorghum and rice. Some historians suggest that women were the first rice cultivators because they were also the main gatherers. They fished at the river's edge, where rice grew. Other historians maintain that rice began as an upland crop. Nature, climate, human usage and migration directed rice's move towards water.

Rice was farmed like other plants, in a forest clearing. Shifting cultivation—called swidden, or slash-and-burn, agriculture—is still used in some upland areas of Southeast Asia and western Africa. Higher yields were obtained through controlled irrigation. Yield intensification produced many innovations, including 'puddling', which creates a belowground level of flat, hard earth. This prevents water from draining away quickly, and breaks down the internal soil structure. Young rice plants gain a foothold over weeds. This method extended use of a limited water supply. Nursery seedlings are transplanted to standing water after one to six weeks. Puddling was probably developed in India, refined and expanded in China, and is used all over the world today.

Other cultivars evolved in the warm, wet and humid areas of southern Asia and the subcontinent, and adapted to differing terrains: lowland rice evolved near rivers and estuaries; upland rice grew on flatlands and mountain slopes in drier and colder climes. In floodplains, rice varieties evolved that could withstand rising waters, with the rice panicle (branching flower clusters with rice grains) remaining above the water.

China, Mostly

The history of rice begins in the foothills of the Himalayas in northeastern India, in Southeast Asia, southern China and Indonesia. Domestication evolved in India and China and subsequently arrived in Korea, Japan, the Philippines, Sri Lanka and Indonesia. Archaeologists have discovered carbonized rice grains pressed into pottery shards in the southern Yangtzi River Valley in China, in Spirit Cave in Thailand, in Koldihwa in Uttar Pradesh, India, and in Sorori, Korea. Fossilized beetles that ate stored rice near carbonized rice grains were also found. The earliest grains, to date, were growing some 15,000 years ago. And glutinous rice, cooked into a thick porridge and mixed with lime and sand, was used as mortar for the 10-kg bricks that make up the Great Wall of China.

Boiling and/or steaming rice is the fastest, easiest way to make it edible. If the grains are soaked first, they get a head start in absorption and remain supple so they don't break when being stirred during cooking. However, if you have many broken grains, you make porridge. Or, if you pound the grains in a pestle and mortar (which you also do to remove the bran covering), you can make flat breads or noodles. Rice paper was a food 'wrapper', especially in Southeast Asia and southwest China. (These wrappers should not be confused with the rice paper that is used by painters or employed for other non-edible functions: edible rice paper wrappers are made from rice flour and water; sometimes egg is added to create dough; non-edible rice paper is made from the straw left on the rice plant after harvesting.) Rice stalks became baskets, mats and sandals, or were burned as fuel. Husks and rice bran oil were used as animal feed, although today rice bran oil is found at upscale supermarkets for cooking and salad dressings.

Beginning in the sixth century CE, southern China became the rice granary for the country; northern China was the political centre, concerned with guarding the border and maintaining hegemony. Rice fed soldiers and prevented famines. *Congee*, a boiled rice porridge or gruel, was then and is still a common meal, even in the north, where wheat and millet were eaten extensively. Called *juk* in Korea, *kanji* in India and *okayu* in Japan, *congee* is found all over the world. One thousand years later, we have Italian risotto and American Cream of Rice cereal: not so different from congee… and yet, both of these soupy rice preparations are unlike congee because of the specific rice and cooking methods used, the function of rice at the meal and consumer demographics. One might say that rice puddings are congees that have been sweetened and stiffened.

During the Song Dynasty (960–1279 CE), Champa rice from present-day Vietnam came to southern China. This rapid-ripening and drought-resistant rice quickly became dominant because it was possible to yield two or three crops each planting season. 'Rice men', specialist rice farmers called *nongshi*, were encouraged to innovate, both in cultivation methods and technology. These men were literate and implemented government policy as they went from village to village. Champa rice provided reliable rice supplies for the farmer, and the second crop was his—after he paid the landlord and taxes—along with the first. The government supported knowledge of rice varieties that could withstand differences in climate, altitude and soil, as well as provide high yields. Rice farmers were skilled in the particular rice cultivars whose seeds they saved, planted, harvested, ate, stored or traded. Hybrids of rice varietals were common, as farmers saved seeds that provided rice plants with desirable characteristics. As a result, rice agriculture became dominant and varied, and when more land was needed, people moved to look for more exploitable terrain. When Marco Polo travelled to China from 1275 to 1292 CE after the Mongolian conquest, he was served rice wine at the palace of the Kublai Khan at Cathay.

For the tables of the Chinese Emperor and his Court, however, white rice was a luxury product. According to one account, a period of culinary excellence occurred during the reign of Emperor Qianlong (1736–1795). Court banquets included soups, fish, meat, vegetables, noodles and sweets. White rice might be served to accompany some dishes, but often it was part of the recipe: lotus root stuffed with glutinous rice, or rice powder dumplings. A rice porridge called *chou* was taken after meals, presumably to aid digestion, since rice porridge was a standard treatment for stomach upset (which is still recommended today) and overeating at banquets was the norm. A seasonal example of *chou* is the La Pa ('harvest') Festival porridge with fruits, nuts and beans.

In 2010, world rice production provided 20 per cent of all calories consumed. One-third was contributed by China, which also fed 25 per cent of the world's population from just 7 per cent of its arable land. Since only between 5 and 10 per cent of the rice produced in the world is traded at all—most is consumed locally—any fluctuations in price, due to climate or politics, has a disproportionate influence on rice in the global market. As urbanization continues, skilled rice farmers are moving to cities, and while mechanization is having some influence on present-day rice agriculture, the loss of skilled rice farmers is worrisome.

Trade and Asia

Within and outside China, rice travelled on barges along canals. The Grand Canal, which was 1,776 km long and transported goods from Hangzhou in the south to Beijing in the north, brought rice to the military. Rice also moved along various 'silk' roads, via donkey and camel caravans. The famous Silk Road, which we are accustomed to hearing about, ran from Central Asia to the Persian Gulf and Mediterranean coasts. The Southwest Silk Road began in Sichuan Province and crossed present-day India (first known as Bharat and later as Hindustan) ending in Bactria, a central Asian kingdom. And finally the Maritime Silk Road led to several ports: Jiaozhou (present-day North Vietnam) and Guangzhou. The Maritime Silk Road wrapped around Indochinese coasts through the straits of Malacca, ending in the Indian Ocean and Persian Gulf.

Of course, ships carried a lot more weight than camels, especially when luxury goods such as spices, furs, ceramics and textiles were being transported. Each transport method had advantages and drawbacks, so all were used. Animals must be fed and maintained, and may become ill. Barges are very slow. Sailing ships rely on prevailing winds ('trade' winds), and often sink, whether from storms or pirate attacks.

Rice was traded for tin (important as an alloy ingredient, as well as a coating for more toxic metals), almonds, wood, ceramics, cowries, ivory, incense and spices. Rice was also used as currency. Certain weights in rice were used as evaluative measures for barter, even when the rice itself did not change hands. It was the gold standard of its time. Rice was also used as ballast on sailing ships, and then sold when the ships arrived in port. Aged rice was considered tastier than fresh rice, so it had an edge over competing grains like wheat, which travelled less hardily.

Arab or Indian traders (Muslims) are thought to have brought rice noodles to Indonesia and the Malay Peninsula as early as the thirteenth century CE. Buddhism, Hinduism and Islam influenced rice cookery. Rice was the basis of most meals, and accompanied the meats, fish and vegetables that were and still are served as curries with *sambals*, spicy condiments made with shrimp paste and tamarind juice. Javanica rice (from Java) is used for curries, while desserts use glutinous rice.

In the 1600s, China set up trading centres on the western coast of Malaysia, and Chinese migrants went to Indonesia. Warehouses and docking facilities had been established in the Strait of Malacca centuries earlier. These narrow protected waters allowed Chinese, Indian and Arabian Gulf ships to meet and trade. The men who travelled to these coasts often married the Malay and Indonesian women they met there. Known as *nonyas*, an honorary title, Malay wives blended their own culinary backgrounds with those of the Chinese. Chinese recipes, wok cooking, spices and ingredients from the Malay community fused into a melting-pot cuisine. Malaysian cookery continues to link four heritages: Malay, Chinese, Indian and Nyonya. Hot chillies and *umami*-rich condiments such as fermented shrimp paste offset the mildness of rice and the richness of coconut milk. Since Muslims are the majority population in Malaysia and do not eat pork, while the Chinese do, and Hindus do not eat beef, they are all linked by their love of rice. In *ikan briani*, a layered rice and spiced fish dish, one suspects the influence of Moghuls from Afghanistan and northern India in the biryani layering technique. Fish fillets seasoned with onions, garlic, ginger, coriander and cumin alternate with layers of spiced long-grain rice seasoned with ghee, cardamom, cloves and cinnamon, the whole finished with chopped ripe tomatoes and thick coconut milk, you could eat this meal. Other dishes were served with unadorned steamed rice, with a spiced accompaniment, such as a fish curry or vinegared chicken. As the Sumatran cookery and foodways writer Sri Owen has noted, rice-focused countries tend to serve rice plain, with everything else on the side.

Another example of *nonya* cooking is the dessert *pulut hitam*, made with glutinous black rice, jaggery, pandan leaves and coconut milk: served with sliced ripe bananas and thick coconut milk, its resemblance to rice pudding is clear, even if the rice is sticky and does not require eggs to bind it, and the jaggery, pandan leaves and coconut milk are replaced in the American/European cupboard by white rice, white sugar and dairy milk, whether fresh or condensed. Instead of ripe bananas and coconut milk, an American version could be made with raisins, while a French version might include candied fruits and nuts, and an Italian version, glazed chestnuts.

A popular Malaysian breakfast is *lakhsha*, a fiery blend of rice noodles mixed with chilli-infused coconut milk, dried shrimp paste, chicken, lemongrass and coriander. The word *lakhsha* comes from the Persian word for noodle. Persians introduced noodle-making to China during the Han Dynasty (206 BCE—220 CE), so it is thought.

Indian Influence

Rice may have been domesticated independently in Afghanistan and northern India at least 5,000 years ago. It spread west to the Indus Valley and south into peninsular India. Rice cultivation began near the Ganges around 2500 BCE. Semi- nomadic hunters and fishermen moved regularly to avoid Mongol invaders from Central Asia, and to find arable land. Around 2000 BCE, these Indo-Aryans moved into the Caucasus, Persia and the Hindu Kush Mountains, settling in Punjab, Delhi and Afghanistan. The five rivers of the Punjab irrigate much of the rice grown in India, even though rice consumption in southern India is greater than in the north. From the Moghal influence comes *pilaus* with cream, fruit and nuts in meat and rice preparation. *Idlis* and *dosas*—both fermented products (souring and fermentation lowered bacterial contamination risk and extended 'shelf life')—and rice and dal (lentils) are more common as you head south. For Kashmiris, pilaf is seasoned with cumin, cloves, cinnamon and cardamom. For Bengalis, the 'holy trinity' flavour profile includes fish, rice and mustard seed oil, just as celery, green onions and green peppers are the 'holy trinity' of Cajun cookery, and *mirepoix* (carrots, celery and onions) for French cookery. For Keralans, curry leaves and coconut perfume rice. Mutton curry with rice showcases the Muslim influence.

When Europeans began colonizing Asia in the late Middle Ages, rice was known as 'batty' in southern India and as 'paddy' in the north. Both terms come from Sanskrit 'Bhakta', which refers to boiled rice. Families farmed small plots of land, producing high yields. In a good year, if the rice deities were pleased and smiled favourably on the family (or if the land owner was pleased with his rice payments), enough rice was produced to provide for the family's immediate needs, with excess rice stored or traded.

Rice can be stored as seed (still in the husk) or as white rice, but not easily in between. The bran and the germ/embryo each contain fat, which quickly becomes rancid in tropical climates. Removing the bran layer and the germ, using a pestle and mortar, results in whitish rice. There are multiple bran layers, so some bran will normally remain unless the rice is completely polished, at which point it becomes white rice and can be stored for years. For some, rice that has been stored for a long time has improved flavour and cooking ability: the grains fluff better after cooking because they have dried out more.

Indica rice is thought to have moved from the Indian

sub-continent and Southeast Asia into Sri Lanka, Malaysia, Indonesia and China south of the Yangtze River. Javanica rice became the upland rice of Southeast Asia and the lowland rice of Indonesia, from which it spread to the Philippines, China and Japan.

India also produces flaked and puffed rice, used in snacks, breakfast and religious ceremonies. Flaked rice can be eaten raw or with milk and sugar (this sounds similar to Rice Krispies, although that rice *is* puffed). Puffed rice is often found in *chaat*, a snack with many variations. *Bhel puri*, a popular street snack in India and in the Indian diaspora, includes puffed rice, potatoes, tomatoes, mint chutney, crunchy gram-flour threads, peanuts, lemon juice, chillies and coriander. A total of 10 per cent of all Indian rice production is utilized for flaked, expanded and popped rice.

Islamic Influence

As early as 1000 BCE, rice spread to the Middle East from India via Afghanistan and Persia. Around 500 BCE, the Persian Empire under the Achmaenids incorporated Asia as far as the Indus River, including Greece, North Africa, Egypt and Libya. Persian influence moved to India via Iran and Afghanistan, spread by Muslim Arab traders, who travelled on to North Africa, Turkey and Greece, arriving in Italian ports, especially Venice and, later, Spain.

Invading Moors from northwestern Africa colonized

Southern Europe and brought rice to Sicily, Southern Spain and North Africa in the tenth century. By the mid-1400s, commercial rice production had begun in northern Italy. Islamic gastronomy was centred in the Abbasid capital of Baghdad, then a cultural and culinary melting pot for the Islamic world, whose agriculture, ingredients and foodways travelled to Spain. The Spanish city of Cordoba became an Islamic cultural and gastronomic capital. Olives, limes, capers, aubergine and rose petals, as well as apricots, artichokes, carob, saffron, sugar, jujube, citrus fruits and carrots, were used in cookery. A condiment similar to fish sauce was made from barley (*murri*) or fish. Long-grain rice was used to make pilaus of all kinds (*pilaf, pilau* and *pulao* are variations of the same word); the paellas, which use short- or medium-grain rice, might be seen as distant cousins. Rice was stuffed into fruits, vegetables, vine leaves and sausages.

Alexander the Great brought rice to Greece from India in the early fourth century BCE. An expensive luxury product, rice was used primarily for medicinal purposes but was also found occasionally at banquets. The Greek physicians Galen and Anthimus both recommended rice gruel with goat's milk for stomach problems, and the rice had to be well cooked. Here is a prescription for an upset stomach from Anthimus:

> Boil rice in fresh water. When it is properly cooked, strain off the water and add goat's milk. Put the pot on the flame and cook slowly until it becomes a solid mass. It is

eaten like this, but not cold, but without salt and oil.

In the seventh century CE, Muslim traders brought Asiatic rice to the Mediterranean. While Muslim Arabs had been trading with China and smaller Asian nations for centuries, they expanded their horizons by turning their focus to Mediterranean cultures and adapted their agricultural methods to Spain, Sicily and Italy, Egypt and Syria. In Egypt, rice was planted near the Nile: the irrigation and transport benefits of the river made Egypt a transport hub for the rice trade. The water-wheel system, known in Roman, Persian, Chinese and Arabian agriculture systems, promoted rice agriculture in the Valencia region of Spain, in Sicily and eventually in the Pô Valley in northern Italy. First built by Romans, norias were giant water wheels. Buckets attached to the wheels' rims deposited water into large and small irrigation canals. More than 8,000 norias were built in Spain. The remains of these water wheels can still be found throughout Spain today.

Chapter 3
The New World

African rice (*Oryza glaberrima*) had been cultivated for thousands of years, primarily in West African coastal countries, in Central Africa and in Madagascar. Red in colour, this rice originates from the Niger River delta, and is the other leading rice family after *O. sativa*, Asian rice. This African staple has significant characteristics distinguishing it from Asian rice. It should be noted that, until the late twentieth century, African rice had not been a major object of academic study. Carl Linnaeus did not include it in his botanical classifications. And wilful ignorance of the importance of African agriculture also accounts for delays in researching its history. In the last 30 years, renewed interest has generated much new research and information on African rice. A comparison of the two plants is helpful.

African rice is somewhat salt-tolerant, a useful trait for irrigation near seawater. The grain is dark red and smaller and nuttier in flavour than Asian rice. It grows vertically without drooping, and is more easily harvested than its Asian counterpart, which is top-heavy when ripe. However, in a pestle and mortar, African grains shatter more easily than

Asian grains; finesse, skill and practice are needed to obtain unbroken grains. Along with sorghum, millets, yams, okra and other plants, African rice was a staple plant.

Both rice and slaves were taken to the colonies of the British, Portuguese, French and Spanish empires. Africans from the 'Rice Coast' countries of West Africa, including Gambia, Angola, Guinea, Guinea-Bissau, Sierra Leone and southern Senegal, as well as the Niger River delta, were sought after for their rice-agriculture expertise. A rice trade triangle developed between Great Britain, West Africa and Western Europe.

Once African rice was planted in the New World, it became critical in the early development of the colonial rice industry, although there was an eventual switch from African to Asian rice for commerce. The African influence on New World 'riziculture' and culinary evolution was extensive. Both Asian and African rice have their advantages and drawbacks in the field, in the milling process and in the kitchen. In the late twentieth century, a hybrid of both kinds of rice, called NERICA (New Rice for Africa), was developed, and has shown promise for the future of African rice. Not only can this rice be grown on upland terrains, it is more pest-resistant and has higher yields than traditional rice, as well as a shorter maturation time: 90–100 days as opposed to 120–40 days.

The British Colonies and South Carolina

British traders who owned sugar plantations in Barbados,

Jamaica and the West Indies discovered that South Carolina's rivers, streams, marshes, tidal basins and subtropical climate were ideal for rice growing. Demand in Europe for high-quality long-grain rice was high. Traders knew that rice would be profitable.

Around 1685, it is said, the physician and botanist Dr Henry Woodward received a packet of seed rice from a ship's captain, temporarily marooned in Charleston. This was *O. glaberrima*, African red rice. Other stories maintain that women and child slaves working in the colonies smuggled in unhusked rice in their hair. Thomas Jefferson even brought Italian rice into the area, as he explained in a letter to the politician Edward Rutledge in 1787:

> I determined to take enough [rice] to put you in seed: they informed me however its exportation in the husk was prohibited; so I could only bring off as much as my coat and surtout pockets would hold.

British plantation owners knew that the Jola, Yoruba, Igbo and Mande peoples were renowned for their rice-agriculture skills, from planting seedlings to building canals and dykes. The Portuguese acknowledged their skills as early as the mid-1400s. These Africans received the highest bids at slave auctions. From inserting a rice seed into a moist clod of earth that would not float away during irrigation, to hoeing, transplanting seedlings, weeding, harvesting, pounding, and husking, slaves worked in the fields, creating 'factories in the field', or rather the colonial rice plantations of South Carolina. African slaves were so adept at every part of rice agriculture that the plantation ran as smoothly as a modern factory, and with an equally huge output. Some slaves grew their own rice (despite laws dating from 1714 making it illegal for workers to grow rice for themselves), vegetables and legumes. Some raised chickens and pigs, and foraged, fished and hunted. They added Native American ingredients to the pantry, especially corn and chillies. And they received rations from their owners, including small monthly allotments of salt, sugar and the least favoured parts of pigs.

Unfortunately, African rice grains broke easily in a pestle and mortar, thereby reducing the yield. In addition, male slaves were often required to undertake this work because female slaves, who were more skilled at milling, were often assigned to domestic chores instead.

South Carolina rice was known as 'Carolina Golde' after the golden fields of grain. It had a very high status among other forms of long-grain rice, and the white grains in particular were very expensive and won awards at European agricultural fairs. The grain was also mentioned by name in British cookbooks used in colonial kitchens.

The first American published cookbook that mentions rice is *The Compleat Housewife* by Eliza Smith (1742). The most popular British cookery book of the eighteenth century, *The Art of Cookery, Made Plain and Easy* by Hannah Glasse (1747), was very influential in colonial households. Recipes for puddings, pilaf, soup, curry and pancakes, all using rice, were included. A recipe for the rice dish hoppin' John appears in *The Carolina Housewife*, in 1847, by Sarah Rutledge, a prominent socialite. By including hoppin' John in her cookbook, Rutledge paved the way for cooking by slaves to develop into 'Southern Cooking'. Still, rice was too complex a topic to easily discuss, as this quote from Mrs Beeton's *Book of Household Management*, published in 1861, makes clear:

> Varieties of Rice.—Of the varieties of rice brought to our market, that from Bengal is chiefly of the species denominated cargo rice, and is of a coarse reddish-brown cast, but peculiarly sweet and large-grained; it does not readily separate from the husk, but it is preferred by the natives to all others. Patna rice is more esteemed in Europe, and is of very superior quality; it is small-grained, rather long and wiry, and is remarkably white. The Carolina rice is considered as the best, and is likewise the dearest in London.

The following recipe from J. M. Sanderson's *The Complete Cook*, published in 1846, shows the early influence of French and British cuisine on the use of rice in cooking, and may possibly have come from a Middle Ages 'receipt' known as 'blanc manger'. This was likely derived from an Arab dish of almonds pounded into 'milk', and rice or rice flour and sugar, enhanced with shredded chicken and rosewater. This pudding has many variants in cooking, and in nomenclature.

Rice Custard. — Take a cup of whole Carolina rice, and seven cups of milk; boil it, by placing the pan in water, which must never be allowed to go off the boil until it thickens; then sweeten it, and add an ounce of sweet almonds pounded.

Some female slaves were assigned to 'the big house' kitchens. The 'mistresses' read recipes aloud from British cookbooks, expecting their cooks to prepare recipes by memorizing ingredients and methods. Despite the European origins of these recipes, African ingredients and cooking techniques (frying, for example) influenced the resulting preparations. Yams, aubergine (eggplant), okra, black—eyed beans (peas), millet, greens, watermelon, squash, sesame seeds, sweet potatoes, kola nuts and sorghum all have an African pedigree.

While choice parts of the pig were reserved for plantation owners and their families, slaves used the hocks, heads, offal, ribs, salt pork, bacon and chitterlings as seasonings or accompaniments to meals, many of which had rice as the centrepiece. Today, these ingredients and recipes remain integral to southern American foodways, not only in black households, but some white ones as well. Rice cakes with honey were occasionally prepared, a Muslim tradition hidden from plantation owners. Senegalese and Nigerian slaves, usually Muslim, cooked their version of hoppin' John with jerked beef because of the prohibition on eating pork.

Descendants of slaves from the coastal lowlands and

Sea Islands of South Carolina and Georgia, called Gullahs or Geechees, used rice to underpin all of their meals. The word Gullah refers to 'the people who eat rice for dinner'. Oysters, shrimp, fish, pork and poultry were occasional accompaniments to white rice. Rice and greens, and rice and okra, similar to Sierra Leone's *plasas* and rice and okra soup, were more common. Red rice, when served with a gumbo (from the Bantu word *nkombo* for 'okra') containing okra, fish, tomatoes and hot peppers, resembles West African jollof rice, which has been described as a 'typical South Carolina meal' and is still typical of Gullah foodways.

By 1690, rice was a substantial commercial crop in the colonies. The speed at which the industry grew is a distinctive feature of Carolina rice culture. In 1700, 600,000 pounds of rice left Charleston harbour for Britain and the West Indies. By 1710, 1.5 million pounds of rice was being exported and by 1740, 43 million pounds. The volume was so great that rice ships became targets for pirates, who ransacked and sank them as they crossed the Atlantic. Rice exports became safer and larger after protective British warships appeared. In 1771, approximately 60 million pounds of rice was channelled through England, which collected duties, to the rest of Europe. Exports reached 80 million pounds by 1789. Fortunes were made. The chart below shows rice price increases from 1772 to 1809.

Rice prices in the USA, 1772–1809

Time Period	Rice Price (cent/lb)
1722–9	1.40
1730–39	1.64
1740–49	1.18
1750–59	1.56
1760–69	1.58
1770–79	1.87
1780–89	3.15
1790–99	2.73
1800–1809	3.81

In 1750, McKewn Johnstone, a South Carolina planter, developed a way to use ocean tides to flood rice fields: this increased available land. And in 1767, Jonathan Lucas invented a water-powered mill, reducing labour requirements and producing more whole grains.

During the Civil War (1861–5), plantations were decimated. The rice industry had already begun a westward move to the prairies of Arkansas, Louisiana and Texas because flat land and machinery (rather than human labour) lowered production costs. In the meantime, chitterlings ('chitlins'), collard greens, ham hocks, grits, black-eyed peas and rice travelled with the escaped and freed slaves as they moved on to northern cities. African American populations grew in Detroit, Chicago, New York and in other cities.

In 1964, the term 'soul food' was used for the first time in print to describe these subsistence food ingredients and preparations that were eaten in a spirit of ethnic pride and 'soulfulness'.

Although rice cultivation continued in South Carolina for 80 years, production was minimal, and ceased around 1920. In 2000, Anson Mills harvested its first crop of heirloom rice, and is now growing other heirloom rices. The Carolina Gold Rice Foundation in Charleston, South Carolina, announced the arrival of aromatic long-grain rice in February 2011. Both Carolina Gold and Charleston Gold are now grown.

Louisiana, the Prairie and California

As the French and Spanish vied for control of the Louisiana territories, rice cookery evolved. The words 'Creole' and 'Cajun' refer to people, ingredients and foodways resulting from European, Caribbean and African cultural and culinary mixing. Creoles began as children born to Spanish parents in Spain's older colonies. Children of French settlers and slaves born in Louisiana were added to this definition. Cajuns were French colonists evicted from Acadia in 1755 when the British gained control of Canada. Acadians ('Cajuns') drifted south and many landed in the bayous of New Orleans. Creole cookery is usually linked with wealthier New Orleans, while Cajun cookery became known a bit later, as the bayous were harder to reach, the people poorer and the food often spicier.

Gumbos are Cajun and Creole rice preparations where the rice serves different purposes. Each recipe may be either Creole or Cajun in origin, depending on whom you talk to. Feelings about gumbo's origins and proper ingredients are fierce. In Creole gumbo, you'll likely find shrimp, sausage and chicken, with rice moderating the spicy flavours and stretching the economic value of the meal. In Cajun gumbo, crawfish, crabs, possibly squirrel, andouille sausage and hot pepper might accompany the rice. Gumbo is a dish in which all the flavours are 'married' to one other; it is remarkably similar to a pilaf or paella, although is much soupier in consistency.

Another iconic Louisianan dish is 'red beans and rice'; a modification of hoppin' John that has just as many variations. Regional differences abound: annatto is added in Jamaica and coconut milk in the Dominican Republic. Hoppin' John is traditionally eaten on New Year's Day, red beans and rice on laundry day. Beans simmered with a ham hock were eaten with white rice. Greens were traditional accompaniments, with the 'pot likker' (the ambrosial water at the bottom of the greens pot) drunk as well.

When Italians, Germans and Yugoslavs immigrated to Louisiana, they added their own touches to the Creole/Cajun melting pot. Italians from Northern Italy had been immigrating to Louisiana since the time of the Spanish explorer Hernándo de Soto (c. 1496–1542), and Germans had come to New Orleans since its founding in 1718. Many Germans moved north to flatter land to become wheat, and later, rice farmers. The German influence is evident in a salad that uses the same seasonings as a German hot potato salad, with a couple of New World additions: bacon, sugar, cider vinegar, celery seed, chopped pimientos, green pepper and onion, all tossed with hot cooked rice, with a sliced hard-boiled egg on top. Boudin blanc reflects the combination of tradition with new ingredients: a sausage of ground pork, rice and spices, it combines the adaptation of German sausage-making traditions with Louisiana's agricultural products, thrift and taste preferences.

In the late 1800s, a farmer planted three acres of rice in Arkansas. The land was flat and could support retooled heavy machinery (originally used for wheat). Improvements in milling and irrigation methods quickly followed. Arkansas became the 'rice basket' of the U.S., as noted in the state anthem: 'There the rice fields are full.' Settlers in covered wagons brought rice to Texas and Missouri, while the Louisiana Western Railroad linking Orange, Texas, to Lafayette, Louisiana, in 1881, brought Midwestern wheat farmers to the prairies of Louisiana, Arkansas and Texas. In just 30 years (from 1880 to 1910) this virgin land, which a government assessor had declared to be of no agricultural value, became the most prolific rice area in the nation.

Midwesterners had to learn how to make fluffy rice. The Southern Rice Growers Association published educational pamphlets with recipes. In 1921, the Creole Mammy Rice

Recipe booklet had recipes for gumbo, jambalaya and rice custard. Also, the nutritional value of rice was emphasized to potato-loving Germans, Czechs and Scandinavians. 'Rice and legumes' were better for you than 'wheat and meat'.

Contemporary American rice production is centred in Arkansas, California, Louisiana, Mississippi, Missouri and Texas. Most rice is long-grain, with Arkansas the leader in tonnage. Production has grown by 60 per cent in the last 30 years, on the same land. In 2010, U.S. rice production was 83 million kg of long-grain, 26 million kg of medium-grain and just under 1.5 million kg of short-grain rice.

Farmers near Sacramento found that rice would grow better than other crops. By 1920, California was expanding into serious commercial rice production; modern 'riziculture' methods include planes spraying seeds, pesticides and fertilizers directly into the furrows of laser-levelled fields. Some 90 per cent of California rice is medium-grain, often referred to as 'Calrose'. It is used for everything from sushi to paella to Asian recipes. It has a clean, mild taste, and sticks together very slightly. In addition to Calrose, other kinds of rice are planted or imported, from organically grown short- and long-grain rices to older breeds like black 'forbidden rice', bamboo green rice and Bhutanese red rice. Organic rice farms have also become an important part of the California rice industry.

In 1919, 1 million acres were used for rice production; the yield was approximately 1,100 lbs per acre. In 2010, 3 million acres were used; the highest yields were 6,500 lbs per acre. In 1970, the rice crop was valued at $0.5 billion, and in 2010, $3 billion. Rice production is divided equally between home consumption and export.

Spain, Peru and Cuba

Between 1849 and 1874, 100,000 Chinese hired labourers, known as 'coolies', came to the Spanish colonies based in Cuba and Peru under eight-year contracts. Most worked on plantations, on coastal farms or as domestics. When Peru and Cuba became independent in the mid-1850s, slavery was illegal and Chinese coolies had replaced slaves as indentured labourers. They demanded rice as part of their payment. Asian rice was imported at first, and later planted along coastal waterways. In the 1870s, escaped and freed coolies moved into the Peruvian Amazon, where they introduced rice, beans and other crops. In the mid-twentieth century, Chinese Peruvians congregated around the central market square in Lima that became known as *el barrio chino* (Chinatown). Today, Chinese restaurants, *chifas*, are evidence of the Chinese–Peruvian mix.

The first Chinese arrived in Cuba in 1857 and worked with African slaves and the indigenous people. Marriages between Chinese and Africans occurred (marriage to Spaniards was prohibited). A total of 125,000 'coolies' eventually came to Cuba. African and Chinese rice cookery combined, resulting in some fusion dishes, while Cuban-

Chinese cooking emerged from what became one of the largest groupings of Chinese in Latin America: *el barrio chino*. *Congee*, the southern Chinese breakfast staple, and *moros y christianos*—black beans and rice (derisively known as 'Moors and Christians')—were both eaten in Havana. After the Cuban revolution in 1959, most Chinese left for Miami and New York, where Cuban-Chinese food abounds.

Mexico

In the 1520s, Spanish conquerors introduced rice to Veracruz, Mexico. Proximity to the Gulf of Mexico made it an excellent choice for rice agriculture. In Campeche two rice crops per season were possible. Rice was introduced into the daily diet, replacing saffron with tomatoes, often cooking rice first in fat and then adding hot liquid and other ingredients.

Cooking rice in fat first proved important to maintaining grain separation after cooking, because the rice was medium-grain, which tends to be somewhat sticky. The Arab/Spanish technique of cooking rice first in fat and then adding hot liquid moved to the colonies in tandem with the advance of colonial powers. Both long- and short-grain rice were planted. Asian rice came via the Pacific route, African rice via the Spanish. Asian rice eventually reigned, becoming the most frequently grown of the two grains.

The traditional duo of rice and beans did not advance into its current iconic state very rapidly. According to

Rick Bayless (chef and historian of Mexican cookery and foodways) rice was adopted more fully once it was incorporated into meals that were already traditional. Usually made with medium- or long- grain rice, *sopa seca* refers to dry rice preparations where all the cooking liquid is absorbed, as in rice and seafood preparations from Veracruz. *Sopas aguadas*, literally 'wet soups', are a stew or thick soup. For the sweet-toothed there is *arroz con leche*, Mexican rice pudding. And for cool relief from the heat, *horchata*, a chilled beverage made of soaked ground rice that is strained and flavoured with almonds or cinnamon.

Portugal and Brazil

Rice cultivation came to Bahia, on the east coast of Brazil, in 1530 when a Dutch ship passed through the Cape Verde Islands with slaves bound for Brazil. From Suriname to Cayenne, stories of slave women and children hiding rice seeds in their hair are legendary. African rice was a provision on slave ships to Brazil. By 1550, the rice was for sale in Rio de Janeiro. By 1618, it was a staple crop for enslaved Africans on sugar plantations in Brazil.

During the seventeenth century, Portuguese and Dutch forces fought for control of Brazil. As plantations developed, more slaves came from Africa. Asian rice arrived in 1766 and was planted for export. By 1781, all the rice consumed in Portugal was grown in Brazil. When Brazil declared independence in 1822, rice production continued as before and became a mainstay of the Brazilian diet. As cities grew, so did rice consumption. This may partially explain why the national dish of Brazil, *feijoada completa*, includes beans, rice, smoked or barbecued meats, kale, orange slices and toasted manioc flour. Manioc is a Brazilian staple, although in recent years its consumption has declined.

Tens of thousands of runaway slaves set up *quilombo* settlements in the dense Maranhâo rainforests. They farmed rice, maize, manioc and bananas and hunted fish and land animals. After slavery was abolished in Brazil in 1888, some slaves left for the cities, while descendants of the original *quilombo* inhabitants still farm rice there today.

When slavery ended legally in Latin America in the late 1800s, new sources of labour were needed for the colonial estates, *haciendas* and plantations, and for building railroads. Organized Asian immigrant convoys to Latin and South America included Chinese, Japanese and East Indian labourers, predominantly men. The Japanese settled in Peru and Brazil, the East Indians went to the British West Indies. Significant numbers of Chinese men left to settle in Brazil, Peru, Cuba and Mexico, where they married local women and adopted their culinary traditions.

'Guinea' (African) rice had been imported to Europe by the end of the fifteenth century by the Portuguese; it was not yet a staple, and more rice was needed. In the 1730s, Portugal imported rice both from Italy and South Carolina, to supplement rising demand for rice in Catholic Europe, which often accompanied fish on holy days (there were at least 100 holy days each year). Persistent cereal shortages in Europe led to rising imports. Brazilian rice seemed to be the answer. Only later was Asian rice planted for export. Rice in Africa was thought to be Asian in origin (even Linnaeus thought this). In Africa rice was brought by Portuguese traders and deposited on the Upper Guinea coast (hence the term 'Guinea rice'). Botanical evidence from the twentieth century confirms that African rice was cultivated at least 4,500 years ago, long before Europeans ventured onto the 'dark continent'. This red rice grew in French West Africa and was an independent species, despite the Eurocentric view that Europe brought rice culture to West Africa and the Americas. When 'Carolina White' rice arrived in Portugal in 1766, to be shipped to Brazil, Italian rice of African origin had already been shipped to be planted. While husking mills removed the exterior of the rice grain, pestles and mortars were still in use until 1774 because of a shortage of mechanical mills. Birds— a major nuisance—dropped seeds from one rice species onto the fields of another, thus making grain separation almost impossible. African rice was finally outlawed and Asian rice took over.

Britain and India

The British began trading in India in the early seventeenth century, with commerce peaking during the Raj era (1858–

1947). Cross-currents between British and Indian gastronomy moved mainly in the direction of Britain, and were accelerated by Indian immigration. Rice started being introduced into the country by returning Anglo officials as well as the Indian seamen who had remained in Britain. The arrival of Indian restaurants and takeaways, as well as the manufacture of processed Indian foods and their increasing convenience, has made British-Indian foods ubiquitous.

From the mid-1800s onward, Britain was the gateway through which rice was taxed before leaving for European ports. Asian rice became increasingly important after the American Civil War cut off rice exports from former British colonies. Revolutions in Europe increased demand for rice. Some rice stayed in Britain to accommodate Indian immigrants, a continuing trend that now includes recent Chinese, Southeast Asian and African-Caribbean immigrants. By the early twenty century military officials and administrators in India referred to themselves as Anglo-Indians. 'Nabobs' were wealthy British men who served during the Raj and returned home after independence in 1947. They often brought their Indian cooks with them. Indian seamen, especially cooks from Bangladesh, came to Britain in the early twentieth century and opened small restaurants. From curry houses to pubs to Veeraswamy's, an Indian restaurant opened by the Anglo-Indian Edward Palmer in 1926 that is still going strong, British versions of Indian food were easily found. While they were not very similar to the meals eaten in India, rice was the common link, and 'curry rice' became the most popular dish. 'Curry rice' is usually served with white rice, unadorned. Fragrant pilafs and biryanis, cardamom-scented rice pudding and kabobs with saffron rice; these dishes are also part of the Anglo-Indian experience.

Originally intended for their immigrant compatriots, Indian restaurants soon attracted British Raj-nostalgic men and women. As the desire for Indian flavours increased, a facsimile of Indian food became the norm. Chicken tikka masala, a British invention, is perhaps the most well known.

Jars of chutneys, pickles, curry mixes and ready-to-use sauces such as tikka masala and vindaloo are commonly found in supermarkets and department stores today. Rice—from Patna to Bengal to Basmati—is readily available. In the freezer are myriad ready-to-heat-and-eat meals, as well as boil-in-the-bag or microwave meals that can be spooned directly onto hot rice.

The British taste for 'Indian food' does not sufficiently explain its great popularity in Britain. A tradition of using spices during medieval times already existed in Britain (among the wealthy, at least) and may help to explain why strong seasonings were so easily adopted by the British, although this idea is controversial. The number of Indian restaurants in Britain exploded after the Second World War and may represent a consequence of rationing, which did not end until the 1950s. Small restaurants became a pathway through which immigrants sustained themselves while assimilating. According to Colleen Taylor Sen in *Curry: A Global History*

(2009), the word 'curry' should be applied only to dishes developed in British Indian kitchens. So where does rice fit in? Rarely mentioned but always assumed, a mound of steaming white rice enhanced the consumption of Anglo-Indian culture in Britain, grain by grain.

An excellent example of cross-culinary pollination is kedgeree (English spelling). This rice and lentil dish, called *khichri* in Hindi, is everyday fare for millions of Indians. Thought of as too low-class for British tables, it was 'elevated' to higher status by the inclusion of smoked fish and hardboiled eggs, an early indication of the middle-class desire for animal protein. It is still served at elegant British lunches.

Indonesia

From the mid-1400s the Dutch and Portuguese alternately fought for control of the Maluku Islands (formerly the Spice Islands, also called Moluccas), the rest of the Indonesian Islands and Sri Lanka. The Maluku Islands traded in pepper, nutmeg, cloves, mace and ginger, which were immensely profitable goods. The Dutch East India Company was begun in 1602 in order to better control trade: as a result Dutch influence became dominant. As with the British in India, the attempts by the colonizers to distance themselves from the colonized became evident at the table. Rice, vegetables and soup were standard fare for the islands' natives, especially on Java, where there was a strong Dutch presence. The extravagant *rijsttafel* the rice table was 'invented' by the Dutch, who sought to make their dining experience more European (or 'sophisticated') by adding many small plates of cooked and raw foods, sauces and condiments, even fried bananas. The *rijsttafel* was a way to show off one's status, and became a traditional Sunday meal. The connection to Indonesian eating habits was one of scale. For the Dutch, having many accompaniments was the point: many dishes signified status of one's colonizing clan, and rice played the role of both sidekick and palate cleanser. *Rijsttafel* restaurants are common in the Netherlands today and appeal to both Dutch and Indonesian people living there. However, *rijsttafel* is not generally cooked or eaten by Indonesians living in Indonesia.

Chapter 4
The Rise of the Consumer

Rice is all one, but there are many ways of cooking it.

Swahili proverb

By 2050, cities will feed 70 per cent of the world population, which will by then have reached 9 billion. Urbanization changes how people acquire and eat rice. Rice provisioning and consumption habits have changed to accommodate these trends. Whether at home or at work, at food courts, corporate cafeterias or restaurants, convenience is the key to this story. While the growing middle classes show a reduction in their consumption of starchy staples in favour of animal protein, affordable rice takes up more and more shelf space in the supermarket.

From the mega mart to the corner bodega, processed rice products flourish as canned, bagged, boxed, chilled, frozen and microwave-ready items. From white to black to wild (which is not really rice at all, since it does not come from the *Oryza* genus but is a semi-annual aquatic grass, *Zizania aquatica*), partially processed rice products can be the consumer's foundation for urban meal preparation. In this chapter 1 will primarily focus on the consumption of rice in the U.S., but we must remember that each product derives from older, more distant, heritages and from more pastoral preparations, and can be found in other parts of the world as well.

Influential Immigrants

'Spanish rice', which included onions, peppers and tomatoes (the inclusion of other ingredients in the definition is more controversial) was common during the expansion of the U.S. in the second half of the nineteenth century, when Texas, California and the Southwest were annexed, conquered or purchased, and the Mexicans living there became 'Americans' by default. The Mexican diet already included rice, which arrived with the Spanish in the 1520s (the Spanish also brought rice from the Philippines via the Pacific Ocean). Today, you can buy 'Spanish rice' in a box.

Italian immigrants were well established in the U.S. by the early twentieth century. Risotto, once exclusively prepared in Italian homes and later in restaurants, can now be bought in a processed format. American arborio rice is presently grown in California, Missouri and Arkansas. I was recently able to purchase a box of risotto alfredo, inspired by fettuccine alfredo. Made with organic California arborio rice, the preparation is simple and quick, requiring butter and milk. The rest of the ingredients are already in the box: rice,

Parmesan cheese, salt, dairy powders, aromatics, spices and oil. The cooking time is 20 minutes, less than half the time of traditional risotto. Four servings cost $3.49.

The Asian immigrant population in the U.S. will reach 20 million by 2020. Recent arrivals have come from Hong Kong, Taipei, Fujian province in China; the Philippines, Malaysia and South Asia. While immigrants' offspring adapt to a more American diet, including an increased consumption of fast food, rice is still often cooked traditionally at home, particularly for weekend events, celebrations and festivals. There is a mixand-match ethic as assimilation crosses with heritage lines. Asian entries in the 'ready meal' category represent another evolution in the market aisle. Lemon grass and rice noodle soup, or sushi wraps with sticky rice: both represent today's ready preparations. These products are found throughout the U.S., above all in California, New York and Texas: the states with the highest percentage of Asian immigrants.

Transportation and Trade

By the 1850s, the process of shipping goods was faster and less expensive thanks to trains and trucks. Refrigerated transport allowed semi-cooked products to move from point of origin to point of sale. As refrigerated planes and container ships joined the mix of transport options, the 'chill chain'—a method of temperature control that maintains freshness and safety of cooked food products—took rice-based meals from factory to refrigerated display case. As a result, cooked foods had a longer shelf life. Frozen foods improved in quality. Complete meals were part of the supermarket larder, only requiring reheating, if that.

Half of the world's rice is transported to cities. Factories and processing plants are often located in between the harvest country and consumption country. Modern rice factories are sometimes even located on land once formerly used as rice fields. Rice has evolved from a staple for survival into a processed product that responds to the shopper's preference and need for flavour, health, value and convenience. In addition, 40 per cent of American rice is used in beer production. Sake, miso and rice wine vinegar use fermented rice.

In my local supermarket in New York I recently purchased a packet of Tilda Thai Jasmine Rice. The rice is harvested in Thailand, milled and packaged in Britain and sold in the U.S., among other places. Modern trade agreements encourage such cross-country production and distribution because they generate cash for the exporting economy, despite problems with subsidies and international trade agreements. When prices are low, rice may be hoarded until the price increases. Thailand is a major rice exporter, especially of high-value Jasmine rice. Both Thai and American governments charge duties for rice imports, just as the British did during colonial times.

The way workers eat has changed. They no longer

routinely eat lunch at home but bring it to work, or more likely, buy it. Whether in an aluminium container, eco-conscious paper container, plastic clamshell or Chinese takeaway container, the midday meal easily maintains food quality and temperature. Sushi and cold rice salads are ubiquitous: they are even now packaged in disposable bento boxes. Many offices now have microwave ovens in which lunch brought from home, purchased from a halal or Desi street cart, a food truck or a supermarket prepared-foods bar, a corporate cafeteria or in a restaurant, can be cooked or reheated. Delivery is standard. Tiffin 'ready meals' are delivered to Indian software engineers in Los Angeles, Toronto and London. This metal Anglo-Indian tiered lunch box, which usually includes rice, dal and curry, is delivered daily, just as it is in Indian cities. The recipients return yesterday's tiffin container for the next day's delivery. Tiffin containers don't require recycling. Could this be a trend?

Innovations

The invention of canning by French chemist Appert was the winning entry in a contest sponsored by Napoleon in the 1840s for safer and longer-lasting foods that could be transported to war zones. Canned cooked rice was eaten by American military troops, especially when added to soups and minced (ground) beef mixes, during the First World War.

Towards the end of the nineteenth century tinplate,

used to make cans, was needed for military purposes other than those relating to food production, so other methods for preserving foods such as rice had to be developed. In 1879 came the folded paper box, invented by Robert Gair. Cardboard boxes had their advantages: the contents could be kept dry, fresh and had a long shelf life, they were well contained and lightweight. The package design and advertising enticed shoppers.

Innovations in refrigeration increased the food choices available to the consumer. Aseptic packaging and vacuum-sealed pouches continue to improve the texture and flavour of processed rice products. Pasteurization and other food safety processes paved the way for mass-market food production not only for supermarkets, but in meals-ready-to-eat—food in pouches that have their own heating elements—for the military. Schools, hospitals and prisons benefit from packaged, canned, frozen or microwaveable rice products. A signifier of this change lies in the U.S. Army manual. A version from 1906 uses canned rice, while another from 2006 refers to individual meals-ready-to-eat of Mexican rice, fried rice and Santa-Fe-style rice.

The locavore movement notwithstanding, we have evolved from consuming that which is grown nearby to that which is grown and shipped in from another country, with convenience and profit in mind. Rice and rice-based products can be ordered online and transported to anywhere in the world. While we associate these practices with economically developed countries, the same trends appear in urbanizing segments of Asia, South America and South Africa. 'Inclusive' packaged rice products make weekday life simple, and the desire for convenience has spread to newer immigrant groups, who may now save major cooking events only for weekends, holidays and festivals.

It is no surprise to learn that frozen rice meals are being manufactured in countries that have longstanding traditions of eating rice cooked in traditional ways. Competition in the commercial marketplace has countries like Thailand contemplating selling prepared frozen meals for shipment to the U.S. Britain is also considering shipping frozen Indian meals (the British version) to India's supermarkets, including chicken tikka masala. This 'added value' to a traditional export commodity will expand the global reach of prepared rice meals, and may just land on your table, wherever you are.

Some Iconic American Rice Products

In 2006 Americans spent 13 per cent of their income on food. Of this figure more than 40 per cent represents food that is eaten out (in comparison, African and Asian households spend 15–50 per cent of their food budgets on street food and eating out). The need for ready-to-eat or heat-and-eat food has accelerated as urban growth increases because new, smaller apartments are being built with mini kitchens, or communal kitchens on one floor (as in Portland, Oregon) and sometimes no kitchens at all, as in Hong Kong and Bangkok. Hotplates and microwave ovens are there to heat foods that are already completely prepared. And rice is a major player in this evolution.

The storeroom, or larder, was once used to stockpile the vegetables, canned, pickled, salted and dried meat, fish and fruit, which were needed to survive the winter. As cities grew, the 'room' disappeared and the 'store' arose. Most storeroom foods are now purchased, as opposed to being hunted or gathered, as in the past. Apartments with small kitchens and little storage space are the norm in cities, while houses in the suburbs have somewhat larger kitchens and more storage space (great places to store convenient packaged or frozen food).

Women entered the workforce in ever-greater numbers after the Second World War. From 1948 to 1985, women in the labour force rose from 29 to 45 per cent. In supermarkets, which at this time catered primarily to women, marketing strategies proliferated to show how quick and easy it was to prepare rice, since women had less time to cook after they came home from work.

In 1904 puffed rice, 'shot out of cannons', was introduced at the St Louis World's Fair. The botanist and cereal inventor Alexander Pierce Anderson, who staged the stunt, knew that water in the precooked rice grain would turn to steam and expand through pressurized heat, thus puffing out the grain. The first commercially manufactured product in the U.S. using puffed rice, Rice Krispies cereal debuted in 1927. An early radio ad targeting children claimed the cereal would 'snap, crackle and pop' in a bowl of milk, and stay crunchy. Of course, puffed rice had been manufactured in India long before Rice Krispies came into existence, and is one ingredient in snacks of all kinds, including *mamra*, puffed rice seasoned with curry leaves, salt, sugar, turmeric, curry paste and garlic, heated until toasted. Similarly, *poha chevda* is puffed or flaked rice with fennel and sesame seeds, salt, sugar, oil and garlic.

The popular dessert/snack known as Rice Krispie Treats, made with melted marshmallows and butter, was introduced in 1941. Today, ready-to-eat Rice Krispie Treats are found individually packaged, and are sold at newsstands and supermarkets. The commercially prepared version is one of the most requested snacks in the U.S. military. Rice Krispie Treats are sold in the United Kingdom, Europe, Canada and Australia.

In 1942, German chemist Eric Huzenlaub licensed his parboiling process to Gordon Harwell, who was selling 'Uncle Ben's Plantation Rice'. Parboiling involves pre-cooking the grain before polishing and pushes 80 per cent of the nutrients from the bran into the grain's centre, thus making white rice more nutritious. The American military became Harwell's chief customer and, by 1944, 20–30,000 tons of rice per year were produced, all for the war effort. After the war, Uncle Ben's Converted Rice was marketed in the U.S., Canada, Australia and Britain; it became the best-selling rice in the U.S., and is still the leading rice sold. Today there are all kinds of seasoned and pre-cooked Uncle Ben's rice combinations. Once again, precooked and dried rice in various forms has existed in India for generations; it is another old idea reinvented for a new market.

In 1958 Vince DeDomenico marketed an old family recipe of rice and vermicelli pasta, both sautéed in butter before being cooked in chicken broth. Armenian in origin, this recipe had its preparation methods in common with early versions of pilaf (first cooking rice in fat, then simmering in broth). DeDomenico's mass-marketed version involved a precooked dried rice and pasta mix in a box, with dried seasoning replacing the chicken broth. The only addition was water. Because the product was half rice and half pasta, DeDomenico called it Rice-A-Roni ('roni' like macaroni). The first advertising commercial featured San Francisco's cable cars and the famous jingle, 'The San Francisco Treat!'

American rice cakes evolved from their Asian and Indian forebears. China and Japan have innumerable kinds of rice cakes and wafers: soft, hard, thin, thick, sweet, savoury. Whether soft, as in Indian *idlis* or crunchy as in Japanese *senbei*, the exact combination of time, temperature, high starch rice and water may produce a soft wrapping for sweet bean paste, a thin batter that can become a crêpe or a crispy, crunchy cracker. Another popular Japanese rice product is *mochi*, glutinous rice, steamed and pressed into cakes, filled with bean paste or ice cream. *Mochi* is found in freezers all over Japan and anywhere there is a Japanese community in the U.S.

One of the first producers of rice cakes in the U.S.

is the Umeya Rice Cake Company, founded in 1924 by the Japanese Hamano brothers, for their own community. Despite internment of the founders in American detention camps during the Second World War, the company eventually rebounded and supplied shops all over the U.S. with the rice cakes. American rice crackers come in brown, white and mixed versions, are extolled for their low calorific value and high fibre (in the version that uses brown rice) and may be seasoned with Cheddar cheese, soy sauce, sesame or caramel.

Minute Rice was invented by Afghani Ataullah Ozai-Durrani and marketed by General Foods Corporation in 1949. From one product came many additions, including today's individual microwavable cups of cooked rice that take only one minute to cook. These rice meals come with package and online recipe suggestions that reflect the immigrant backgrounds of the American demographic. From curried rice to salsa rice to Asian chicken rice to Greek rice salad, whether you are bringing to life your own heritage or somebody else's, Minute Rice will help.

Instant rice noodle soup in a microwavable cup is advertised to workers, students, parents and the elderly, for a quick, easy, 'no dishes to wash' meal. Wheat and rice noodles were first pan-fried and dried, then added to Styrofoam cups and bowls with dried vegetables and a seasoning packet. Add boiling water and wait a minute or two. Invented in 1958 by Momofuku Ando, this soup in a cup was once voted the most popular Japanese food invention in Japan.

Restaurants

In eateries and restaurants, rice is ever-present. This was not always the case. As immigrants from rice-focused countries have increased in number in the U.S. and Europe, so too have eateries that cater to them, and to the rest of us. 'Authenticity' is not the point, although a close resemblance to 'the real thing' is always a good selling point, no matter how far-fetched the link.

During the 1850s, Chinese workers in California opened small, simple restaurants in their 'Chinatowns' in order to cater to their compatriots. Steamed white rice mounded in a bowl was the backdrop for pork, greens, tofu, seasonings of fermented black beans, soy and oyster sauces, as well as garlic, ginger, spring onions (scallions) and sesame oil. Eventually, menus expanded as non-Chinese customers arrived, although there was often a separation between what the Chinese and non-Chinese ate, with the exception of rice, of course!

What we think of as 'fried rice' began as a way to use up leftovers in the kitchen and was not included on the menu, at first at least. Nevertheless, fried rice in various guises (pork, shrimp, tofu) has become a staple in restaurants today, whether Cantonese or not. Fried rice is now available as Minute Rice, a ready-to-eat microwave meal.

The Asian immigrants that came to New York, California and Texas beginning in the 1970s prepared rice that reflected their backgrounds. Pho—the archetypal Vietnamese rice noodle soup—is found in San Jose and

Houston; Thai sticky rice with mangoes and coconut is eaten in Los Angeles and New York City; Fujianese chicken with fermented red rice paste can be found in Queens, New York and in Little Fuzhou, in Brooklyn, New York. Indonesian, Korean, Singaporean, Chinese and Malaysian rice dishes are also represented.

According to folklore, risotto emerged in Italy in 1574. A stained-glass maker, working on Il Duomo in Milan, added saffron to his glass recipe to achieve a bright yellow colour. Liking the results, as a joke, he also coloured the beef marrow and rice with saffron at a wedding feast, but the guests found the dish delectable. It became known as *risotto alla Milanese*. The classic way to make risotto involves cooking short-grain, high-starch rice in butter or oil, so that the grains are coated with fat. Hot stock or other liquids are added very slowly, and the mixture must be stirred continuously while the starch is slowly released from the grains, which acquire a chewy, creamy texture. Butter and grated Parmigiano-Reggiano cheese are added shortly before consuming. When risotto is featured on menus in Italian restaurants, there is often a caveat: 'The risotto will take 25 minutes to prepare.' In addition to the classic *risotto alla Milanese*, restaurant goers now find all kinds of risotto on the menu. Speedier ways to prepare risotto have evolved, and less traditional ingredients have become standard: vegetarian risotto with tofu, and non-dairy risotto (made with rice milk and puréed nuts for creaminess). Risotto now even comes in a box.

Paella originated in Valencia, Spain. The word 'paella'

may be a corruption of the phrase *por ella*, which means 'for her'. (According to legend, a lover once prepared *por ella* for his fiancée.) Or, it may come from the name of the round, shallow pan with two handles that is used for cooking paella, a *paellera*. Paella began as a mixture of rice, vegetables, rabbit and snails that labourers cooked over an open fire in the fields and ate directly from the pan. A *soccarat* is formed when the bottom of the pan remains very hot and browns the bottom-most rice, which becomes crunchy. In the U.S., *paella Valenciana* has come to mean a preparation that includes chicken, chorizo, shrimp, clams, short- or medium-grain white rice, vegetables and seasonings, and saffron. And yes, paella comes in a box or as a frozen meal.

Pilaf was originally destined for aristocrats. Called *pulao* in Iran, Afghanistan and India, the best pilaf is made with aromatic Basmati rice that is usually aged for several years, and therefore is expensive. The goal of every pilaf cook is to serve fragrant mounds of steaming individual rice grains. To make pilaf, rice is washed or soaked, then drained until the water runs clear of starch. A spice mix of pounded green cardamom pods, cumin seeds and cloves is cooked in ghee with caramelized onions. Rice is stirred into the mixture and the grains are coated with fat. Water or stock is added, and the rice cooks at a simmer until 'steam holes' appear. Then the covered pot steams for another 20 to 30 minutes. Putting a dry towel on top of the rice, inside the pot, helps absorb the water condensing on the inner lid and ensures that the grains will continue to separate from each other. This prevents gummy rice. Afghans and Iranians often pour ghee into the centre of cooked rice where it spreads across the bottom of the pot. As the rice steams for 30 more minutes, a *tahdig* is formed: a browned crunchy crust on the bottom of the pot.

While the origins of gumbo will be forever debated, there are some facts that are not in dispute: they may seem contradictory, but that is part of the story. Gumbo is the essential Creole (for some) or Cajun (for others) soupy rice dish that gets its name from okra (from the Bantu word *nkombo*). The legacy of French colonialism includes roux, a butter or oil and flour mixture used for thickening. Some gumbos use ground sassafras as a thickener; a Choctaw influence. Cajuns use crawfish in their gumbos. Spicy sausages may have come from the Cajuns, and smoked sausage from the Germans. The use of tomatoes depends on the cook and her or his traditions.

The Street

Street food is the most convenient food. Found at stalls and carts, food trucks and outdoor markets or festivals, it has evolved into the 'ethnic' fast food found in food courts, where Chinese, Indian, Thai and Mexican restaurants set up shop. Anything wrapped, stuffed or easy to eat is suitable street food. You can indulge in crispy, sugar-coated Yunnan strawberries wrapped in rice paper in the Chinese city of

Kunming, or gorge on goat biryani in Mumbai in India. How about Korea-town (in New York) *kimbap*: steamed rice, stuffed with pickled radish or tuna, rolled in toasted seaweed sheets? Or *bolinhos de arroz*: rice fritters stuffed with sardines and cheese in Rio de Janeiro? Or in New Orleans *calas*, fried dough balls made with cooked rice and sprinkled with powdered sugar.

Some street foods require more than one's hands. *Pho*, Vietnamese rice noodle soup, Tex-Mex *carne asada* with rice and beans, or Gambian chicken *Yassa* with rice: these foods are derived from Vietnam, Texas and Gambia, but are also found in New York, New Mexico and California, courtesy of the immigrant presence there.

Street food can be an inexpensive and practical way to feed one's family. In southern China, huge pyramids of *chong* pile up in bamboo steamers. Consumers unwrap the banana-leaf-wrapped rice balls filled with seasoned mushrooms and eat them piping hot. Coconut-scented packets of *puso*, jasmine rice wrapped in boiled coconut fronds, fit the bill in Luzon in the Phillippines. In southern India, *idli*, puffy, light-as-air steamed rice-and-lentil cakes, are a common breakfast food, and let's not forget *dosas*, those fermented rice and lentil 'crêpes' that enclose curried vegetables or potatoes with mustard seed, among many variations. In Indonesia, street fare might be *nasi goreng*, fried rice flavoured with *belacan* (shrimp paste) and shredded cooked egg.

Street carts and food trucks abound. They bring their rice meals to busy neighbourhoods, food festivals, farmers' markets and agricultural fairs. These movable feasts offer vegetable kormas with spiced basmati rice, or Korean fried rice with *kim chi* and pork. The Kogi BBQ truck in Los Angeles, which started the food truck movement, has recently opened a restaurant: the reverse of the usual scenario. What is served? Korean rice bowls, with meatballs, pork belly or tofu, and Korean pickles.

Sushi: A Special Scenario

Sometimes, street food is elaborate and expensive; at other times, simple and cheap. Sushi, too, can be either.

It is thought that during the second century CE near the Mekong River, in landlocked parts of China, Laos and northern Thailand, salted fish was placed between layers of cooked rice. The container was sealed for a lengthy period of time, during which the salt, and rice fermentation, preserved the fish. The rice was discarded before consuming the fish, although a few aficionados enjoyed the intense aroma (much like blue cheese). The preparation time and costs involved made this food for the wealthy. For those who lived near the water, acquiring fish was much easier.

This sushi style migrated to China and Japan in the seventh century. The Japanese ate the rice and the fish, while the Chinese did not take to it. Increasing popularity for this rice/fish combination in Japan resulted in shorter preservation times, so as to keep up with demand. While the fish was still protected from bacterial contamination, the rice became much more palatable. In 718 CE, sushi was acceptable to the government as a form of tax payment. By the early seventeenth century, sushi rice was seasoned with a revolutionary new product: vinegar made from rice, 'rice wine vinegar' (which actually has nothing to do with wine). Seasoning the rice with rice wine vinegar permitted fish and rice to be eaten together because the rice wine mimicked the flavour of fermented rice, and had preservative qualities as well. Sushi consumption had also become much less expensive and more democratic, moving from the aristocracy to the labourers who frequented street markets at lunchtime. Simple rice and fish combinations as well as elaborate bento boxes of prepared sushi, with pickles and other condiments, were ready to be eaten anytime.

After the Second World War, sushi stalls and carts in Japan were moved indoors to more sanitary conditions. As Japanese restaurants proliferated, sushi began its global migration while adapting to newer environments, giving rise to the California roll, the inside-out roll and sushi with brown rice. Top sushi chefs are very particular about which rice they use, and some even polish their own rice (top-grade short-or medium-grain rice). Apprentice sushi chefs begin their long training by learning to prepare rice.

Sushi is served from high-end restaurants to institutional cafeterias, and can be purchased at supermarkets and at stores that sell only pre-packaged sushi. This style of sushi is prepared in factories where robots mould the rice into individual oblongs or rolls. Trained staff add sliced fish. The use of frozen farmed fish and lesser-quality rice has brought the price of sushi down from food for the wealthy to food for the masses. As a result, sushi is almost global in its reach. Some sushi restaurants use conveyor belts that move around a circular counter: customers take small plates as they pass by.

Perhaps the best example of the sushi explosion is in São Paulo, Brazil, which has an abundance of sushi restaurants. Japanese immigrants first came to work in this area on coffee plantations in 1908. The Japanese population grew, and today São Paulo has the largest such community outside of Japan. Seventeen million sushi meals are eaten every month. In Brazil, sushi began as a luxury product for wealthy diners. As its popularity increased, automation of rice and fish brought the price down. Cafeterias and salad bars now offer sushi. One change that reflects the background of Brazil's history with rice is the preference for serving long-grain rice with Japanese foods other than sushi. Outside of Asia, Brazil has one of the highest consumptions of rice: 40 kg per person annually. And sushi is often prepared using Brazilian ingredients to reflect local tastes: mango, strawberry and raw beef.

Sake

Fermented rice and water combinations have existed in China and Korea for millennia, but Japanese sake—the one with which we are most familiar—began about 2,500 years ago. Moulds, fermentation and wild yeasts, and the even earlier chew-and-spit technique, were used to ferment brown rice and water into gruels that were strained. The resulting brew was light brown and cloudy. In AD 689, the new brewing department at the Imperial Palace, along with Chinese help, developed moulds that increased alcohol content. The ritualistic aspect of sake drinking solidified when Shinto monasteries became legal breweries in their own right. During the next 400 years, sake brewing became a business. Kyoto and Kobe became major brewing centres. Government taxes ensued. In the late 1500s, when rice began to be polished, sake could be made either cloudy or clear. By the 1800s, just as with single malt scotch, wine and cheese, local conditions—including time of year, rice type, terrain, climate and water—influenced the flavour of sakes, and local sakes came to be preferred by different consumers. During periods of rice shortage (the Second World War, for example), mixing a small amount of rice with distilled cheap alcohol produced low-quality sakes. After the Second World War, whiskies, wine and beer became popular in Japan, while sake became more popular in Europe, South America, Australia and the U.S.

Sake came to Hawaii in 1885 with Japanese labourers on sugar plantations. Import taxes kept increasing, so Japanese-owned breweries were built on American soil. Initially intended to increase sales, lesser-quality sake was produced. This sake was heated, which helped disguise its poor quality.

Other deviations from Japanese etiquette included drinking sake with sushi. Not only does traditional sushi etiquette dictate that sake not be drunk with sushi ('like with like' is frowned upon in some etiquette circles), but high-quality sake should be served chilled, with vintages and other labelling information available to connoisseurs.

The Electric Rice Cooker

Rice is usually simmered, boiled (like pasta), steamed over water or cooked with a combination of these methods. Any of these techniques may result in imperfectly cooked rice, especially for rice cooked in a pot and requiring one to tend to the flame or coal and adjust the rice by moving the pot and rice around. While reliable gas stoves have existed for more than a century, the rice cooker revolutionized rice cooking in Japan, and from there, the world.

The first automatic rice cooker was manufactured by Toshiba in 1955. A rice cooker that kept rice warm all day was introduced in 1960, and became wildly popular. Sushi restaurants also greatly benefited. In 1979 digital technology allowed one to set a timer the night before and have hot rice for breakfast. In 1988, Matsushita introduced induction heat cookers. While slow to catch on, they now comprise more than half of rice cooker sales, despite being more expensive. The advantages include not having to soak the rice and, with a gentle stir at the end (which lets excess moisture evaporate), a more consistent product. In 2003, Matsushita introduced a rice cooker that also uses very hot steam: this makes the rice more aromatic.

Matsushita has found sufficient interest in rice cookers in Europe and the U.S. to produce specialized rice cookers tailored to customers' preferences. In the U.S., a basket above the cooker is included, for steaming vegetables. The lid is transparent so you can see when everything is ready. Rice cookers vary depending on which rice you prefer, whether it is parboiled and whether you steam or boil it. Rice cookers, just like rice, have adapted to the needs of their audience.

Chapter 5
Art, Ritual and Symbolism

Luck is like having a rice dumpling fly into your mouth.

Japanese proverb

Rice has inspired origin myths, customs, rituals, language and attitudes. Sometimes traditions have developed into newer paradigms after people migrate, or when modern technology is implemented. Pesticides replace prayers, and rapid-growth rice supersedes traditional varieties. Modern technology in rice production is also the source of lost traditions and increasing unemployment, as machines replace humans.

Attempts to save on labour with research into direct seeding, mechanical rice transplanters, weedicide screening, and mechanical threshing are conducive to despair as a use of aid funds in Asian research centres. The issue is largely a matter of timing. As economies grow, the point is reached where there is no longer a surplus but a shortage of labour in the agricultural sector ... The unmistakable trend is marked by the gradual disappearance in many regions of practices and techniques that have been used for centuries in the production of rice. The tractor is replacing the water buffalo for land preparation; direct seeding of rice is replacing transplanting; herbicides are replacing hand weeding; the mechanical thresher is replacing traditional hand threshing of the paddy... While the youth no longer look to rice farming as a way of life, those left behind to tend the rice fields are adopting new practices to lighten the burden and increase the productivity of their labours.

Despite the changes discussed above in 'Rice Research and Production in the 21st Century', a report by IRRI in 2001, ancient traditions are valued and rituals are maintained when it comes to rice. Religious shrines are found in rice paddies and religious harvest ceremonies are common. The Water Puppets of Hanoi theatre fame mimic rice planting, and harvest symbolism, in their performances. Followers of Mae Posop, the Thai rice goddess, discreetly cut rice stalks when harvesting them so as not to offend her. Baskets of rice are poured over the heads of Indian brides and grooms. Handfuls of rice are tossed at American weddings as well (although in America birdseed is now thrown instead if the ceremony is held outdoors, since birds cannot digest raw rice). There is some overlap in myth origin stories; this makes sense when you understand the complex way rice moved from people to people, and land to land.

Not all rice imagery is life-affirming, however. Rice grains and rice agriculture have been metaphorically used to negatively denote racial and gender distinctions.

Origin Stories and Myths: Gods and Rice

Dietary staples usually have a variety of origin myths. Frequently, gods are credited with giving or withholding basic foods: wine, beer, corn, chocolate, wheat and rice have all been implicated. To encourage abundant harvests, people worship at shrines and temples, and make offerings to male and female deities, whether human or animal. An abundant rice harvest not only guarantees immediate survival, but is also a hedge against future privation.

Some gods are vengeful and even feel guilt. The Javanese story of Tisnawati is illustrative. Tisnawati, a god's daughter, falls in love with the mortal Jakasudana. Tisnawati's father does not approve of the relationship between god and mortal. As punishment, Tisnawati is changed into a rice stalk. Later, her father takes pity on her lover, and turns him into a rice stalk as well, placing him next to his beloved. Their union is re-enacted at the harvest festival, and symbolizes the triumph of committed love over more temporal and less reliable emotions.

Commitment, a willingness to work very hard and choreographed teamwork are required for a good rice harvest, and these three elements may produce an affinity for mathematical reasoning, as Malcolm Gladwell points out in his book *Outliers: The Story of Success* (2008). While this is a stereotype (hence its inclusion in 'Gods and Myths'), Gladwell points out that rice paddies in Southern China are very small, and the whole family gets involved in all aspects of rice agriculture. The timing of plantings, square footage, land maintenance, water levels, a level claypan, weeding, irrigating and all the other labour-intensive parts of rice farming are still manual. Decisions are made daily and many involve arithmetic and fractions. Add to this the more rapidly developed memory skills of younger Chinese in terms of numbers (compared to Americans), and the hours spent working (both in the paddy and in the classroom and at home) and you have an explanation for Asian success in American mathematics classes. Singapore, China, South Korea and Japan are linked through wet rice agriculture and specialization in mathematics.

Other gods are generous. In Tamil Nadu in India the rice goddess is Ponniamman. The name is an amalgamation of the name of the 'Ponni' rice that grows in the area and the Tamil word for goddess, *amman*. Heavy flooding frequently washed away rice fields in the region. A statue of Ponniamman was erected, prayers were offered and the floods abated.

A Chinese rice-origin myth tells the story of a people who had taken refuge in the mountains after floods destroyed their plants. A dog ran by with rice panicles hanging from its tail (the panicle is part of the rice tiller that bears rice spikelets, which develop into grains). The seeds fell to the ground as the dog ran, and rice shoots grew where the seeds fell.

Also in China, a mythical hunter named Houh Jir had five sons. He gave each one a sack to fill with one of the five staples: hemp, wheat, millet, beans and rice. The son who filled the bag with rice was named Pahdi: this was the origin of rice, and it became known as 'paddy' in honour of the son.

On the rice terraces of the Ifugao in the Philippines, bulol figures, in male and female pairs, represent rice gods. The god Humidhid made four carvings in human form, from the dark wood of narra tree, a national symbol of the Philippines. The statuettes went downriver, where they multiplied and protected the rice paddies and the granaries where rice is stored.

Inari, the Japanese rice god, is thought to have cultivated the first rice plant. While a snake guarded a bale of rice, a fox came along to do his bidding, which included dropping rice seeds to be planted. Families often have small shrines to Inari at home, and Shinto and Buddhist shrines celebrate Inari. Also called the 'God of Prosperity', Inari stands on two bags of rice while riding the fox. There are numerous variants on this theme.

Celebrations: A Rice Festival Sampler

In Bali, the harvest festivals honours the goddess of

rice, Dewi Sri, in three forms: Tiswanati, the goddess who gave birth to rice, and whom we have met before in Java; Mae Posop, the goddess who protects the rice crops (she is also the Thai rice goddess); and finally, harvested rice that appears as a mortal woman. One story explains the origin of the rice plant this way: the god Batara Guru was given an egg-shaped jewel. When he opened it, he found a beautiful young girl, whom he named Tisnawati. All were heartbroken when she died as a young woman. After her burial, the king was riding in the forest. Near Tiswanati's grave, he saw a beautiful shining light. As he came closer, he found that a coconut palm grew from her head, the banana tree from her hands, corn from her teeth, and rice from her genitals.

During festivals, villages are repainted and decorated with flags. Rice-god statuettes dusted with rice flour that honour Dewi Sri are placed throughout the fields. Models of the mother spirit made from rice sheaves are hung near paddies to ensure good crops. Granaries are the dwellings of ancestral spirits, so they are often built to look like small houses, in which ancestors have enough rice to eat until the next harvest. Ceremonial objects, usually made by the senior woman of the household, are placed in the granary. Meals of spit-roasted duck and pig are eaten with *nasi goreng* (fried rice), rice sweets and dumplings.

The state of Tamil Nadu hosts a four-day Hindu harvest called Pongal to give thanks to the sun, rainfall, cattle and cereal grains—rice in particular—as well as for sugarcane and turmeric. On day one, people light a huge bonfire at dawn and throw old and unused items into the fire. They clean their homes and decorate the ground outside the entrances with elaborately coloured rice-powder designs. The farmers anoint their ploughs and sickles with sandalwood paste before using them to cut the newly harvested rice. The sun god, Lord Surya, is honoured on day two. He is offered rice that has been cooked in milk in an earthenware pot and decorated with turmeric and sugarcane. When the pot boils over, devotees share the cooked rice. Buffaloes and cows are washed, garlanded with flowers, and worshipped on day three to give thanks for ploughing the fields. They also eat rice boiled in milk. On day four, young women prepare and leave balls of cooked rice out in the fields for the birds to eat. The Pongal festival is considered propitious for marriages, as those with a plentiful harvest can supply rice and other necessities for the wedding.

The June rice-planting festival in Osaka, Japan, reproduces ancient rituals. The paddies are still tilled by oxen. In larger cities, rice is sometimes planted on the rooftops of skyscrapers. Performing spectacular dances and singing special songs invigorates the grains and reinforces beliefs that spirits live in the rice seedlings, which must be planted in Mother Earth. The women sing and dance while wearing *kasa* (braided hats) festooned with flowers, while a procession of samurai warriors clad in full armour passes by. The highlight is the *Sumiyoshi Odori* dance by 150 young girls. Participants' prayers are answered with the autumn rice harvest. Once harvested, rice offerings are made to the shrine deities in

October and, on 23 November, a ceremony is held to express thanks for the bounty.

In China, the La Ba Rice Porridge Festival, Spring Festival (Chinese New Year), Dragon Boat Festival and Mid-Autumn Festival are only a few of the festivals that celebrate the importance of rice in everyday life: the arrival of new rice, a plentiful harvest, the solstice and, in our example, matchmaking. At the Guizhou Sister Meal Festival, Sister Rice is a culinary and symbolic key. The intensely coloured rice is made by young women who collect leaves, flowers and grass to make coloured water, in which they soak the rice for a few days. They put tokens in the rice and give them to Sister Rice to hand out to young men that catch their eye. The rice colour and the token indicate different interests. Red rice, for example, means that the village from which the woman comes is flourishing. If cotton is found within the rice, the woman is showing her eagerness to marry, if garlic, the opposite is true.

A more contemporary honour takes place in Mengzi County, Yunnan province, in southwest China, and celebrates a dish called Crossing Bridge rice noodles. The story is said to have arisen from a wife's efforts to provide her husband with hot lunches, while he studied for exams on a quiet island in Nanhu Lake, far from home. His wife came with lunch every day, but by the time she crossed the bridge to reach him, the soup was cold. One day, she made chicken soup. When she arrived at her husband's study, the soup was still steaming hot: it had an insulating layer of chicken fat floating on top, and this kept the soup warm enough to last the long trip. As a result, chicken soup became the foundation of the meal, rather than the meal itself. Various thinly sliced, uncooked ingredients, such as rice noodles, meat, fish and vegetables would also accompany the soup. These items would be tossed into the scalding soup just before eating it, ensuring a fresh and hot meal.

In both North and South Korea, noodles (symbolic of long life) and dumplings are often served on birthdays. Steamed glutinous rice cakes are flavoured with chestnuts, honey, jujubes, sorghum and mugwort. Especially favoured in South Korea, rice cakes flavoured with mugwort leaves are supposed to have therapeutic value. These cakes are traditionally served at the Tano festival.

During the early Choson dynasty (1390 to 1910, but especially during the sixteenth and seventeenth centuries) and concomitant rise of Confucianism, holiday rituals and seasonal festivals flourished. Harmony with nature was encouraged through the consumption of rice-cake soup, rice cakes steamed on pine needles, sweet rice beverages and more. Rice-cake soup was eaten on New Year's Day. And on a child's one hundredth day of life, he or she would receive steamed rice cakes, which represent innocence and purity. On the child's first birthday, a layered rainbow rice cake might symbolize his or her future endeavours.

Rice is significant in many West African ceremonies. While women are the primary field workers, various rituals include masquerades performed by men, with masks and headdresses that mimic animals and birds. Feasts that use rice as the centre of the meal support the link between rice and women's fertility. Elaborate woodcarvings and winnowing baskets made from reeds and other grasses are also part of these rituals. The baskets are still made in Africa, and can also be found in the Low Country of South Carolina today.

To promote a good harvest in Liberia and on the Ivory Coast, Dan women dance with wooden rice ladles carved to portray animals and people, while in Mali, Bamana men wear antelope headdresses and dance to the spirits. At Baga weddings in Guinea and Liberia, brides dance with baskets on their heads, into which onlookers toss gifts of rice and money.

In the U.S., Arkansas, Louisiana, South Carolina and Texas all have annual rice festivals. The festival is a time to celebrate the harvest, eat lots of good food made with rice and highlight the importance of the rice industry to the community. The International Rice Festival in Crowley, Louisiana, includes a parade, a ceremony that involves crowning the 'rice king' and 'rice queen' and rice eating and rice cooking contests. Every kind of rice-based food is available, from boudins (a sausage made with rice), to jambalaya to gumbo.

The Texas Rice Festival is an annual harvest celebration held in October, around the town of Winnie, Texas. The event features a carnival and parades, a livestock and longhorn show, a horse show, barbecue cook-off, nightly street dances, rice cooking contest, pageants and features food made with rice and flavours of Cajun culture, which is strong in the area. Typical fare includes rice balls, gumbo, *étouffée* (often crawfish and other shellfish in a dark roux-enhanced sauce, served over rice), crab balls, boudin balls (sausage and rice balls) and funnel cakes, as well as cowboy casserole—a one-pot recipe that ordinarily includes minced (ground) beef, vegetables, beans and tomatoes, with biscuits or cornbread or rice, all cooked at the same time.

Beginning in 1976, the Arkansas Rice Festival included choosing a Miss Arkansas Rice Queen every year. All sorts of rice cooking contests take place and draw famous local chefs to demonstrate their skills. Rice is even eaten with butter and sugar.

Paella is the star of the Fiesta del Arroz (Rice Festival) in Sueca, an agricultural town south of Valencia in Spain. This annual festival occurs during September and celebrates the most internationally known rice dish of Spain. Local and international chefs compete in the 'international paella contest' using the famous local rice, including *La Bomba* and other special short-grain rices that have rounded as well as long grains, and are especially absorptive. Valencia paella is traditionally made with chicken, rabbit, land snails and greens, although every village is proud of its own version. The most popular specialities are paella with chicken or rabbit, seafood paella or a mixed version. Among fishermen, *arroz a banda* evolved, so called because the rice and the fish are cooked separately, to fully develop the flavour of each part. It is served with aioli (garlic mayonnaise). Rice baked in earthenware pots with beetroot (beets), cuttlefish, cauliflower or spinach is also popular.

Cultural Customs and Rice Rituals

Showering brides and grooms with rice is an ancient ritual used by Assyrians, Hebrews and Egyptians. In a Hindu wedding ceremony, the bride's brother pours unhusked rice into the hands of the bride and groom as they walk around a fire. The couple offers this rice to the fire. Rice is the first food a new Indian bride gives to her husband and is also the first solid food eaten by an Indian infant. Rice and fertility are almost synonymous.

At Japanese weddings, rice cakes embossed with cranes or turtles are offered to the bride and groom as a symbol of longevity. Red beans and rice are given as gifts at a child's birth, while at Buddhist funerals, puffed rice is used to symbolize rice that cannot be grown again. At Korean funerals, three spoonfuls of rice are placed in the deceased person's mouth, along with some money, to ease the transition into the next world.

Language, Literature, and Art

The words 'rice', 'food', 'meal' and 'eat' are almost equal in several Asian languages. 'Agriculture' and 'rice' are often used synonymously. In myriad ways, rice is a vehicle of expression.

There are no references to rice in the Bible, but Confucius and Muhammad both considered rice as their favourite food. Muhammad liked rice cooked in ghee (clarified butter), often sweetened. Buddha, when he was still Siddhartha, was a man of many pleasures who turned to asceticism while seeking enlightenment. For many months, he lived on one grain of rice per day. One day, a young girl brought him a dish of rice cooked in milk. Restored, he was able to continue his quest. Buddha had rice-paddy designs sewn into his robes, designs still used today.

Devotees of Krishna were told that foods were divided into three categories, of which the first category includes rice, milk and dairy products, all aligned with virtue.

Many proverbs and sayings feature rice as the vehicle for conveying meaning or sentiment:

> Don't let an angry man wash dishes; don't let a hungry man guard rice.
>
> —*Cambodian proverb*

> If you are planning for a year, sow rice; if you are planning for a decade, plant trees; if you are planning for a lifetime, educate people.
>
> —*Chinese proverb*

> Try to seize the bowl of rice but forget the whole

table of food.

—*Vietnamese proverb*

Rice is born in water and must die in wine.

—*Italian proverb*

Have you eaten rice yet?

—*Traditional Thai greeting*

The great American jazz trumpeter Louis Armstrong's favourite food was red beans and rice. In tribute to this iconic dish from New Orleans, he signed his letters, 'Red beans and ricely yours'.

Another Side of Rice Symbolism

Negative racial stereotypes are embedded in the language of rice. During the Vietnam War, a child born to a Vietnamese woman and an African-American soldier was sometimes described as the 'colour of burnt rice'. In a poster from the same period, South Vietnam used the imagery of an abundant rice harvest combined with rows of marching soldiers, bayonets raised, as propaganda to support Vietnamese soldiers. They would be able to fight well and long, as long as there was rice to sustain them.

The Chinese-American author and educator Maxine Hong Kingston describes her grandfather's attitude toward girls in her book *The Woman Warrior* (1975) in terms of rice: 'The families are glad to be rid of them. Girls are maggots in the rice. It is more profitable to raise geese than daughters.' In the Philippines, *ampao* is the name of a puffed rice delicacy that, when used pejoratively, refers to someone with an empty head.

In *The Member of the Wedding* (1946), Carson McCullers describes the main character's love of rice:

> Now hopping-john was F. Jasmine's very favorite food. She had always warned them to wave a plate of rice and peas before her nose when she was in her coffin, to make certain there was no mistake; for if a breath of life was left in her, she would sit up and eat, but if she smelled the hopping-john and did not stir, then they could just nail down the coffin and be certain she was truly dead.

Rice in this Filipino song is a paean to hard physical work and drudgery:

> Planting rice is never fun;
> Bent from morn till set of sun;
> Cannot stand and cannot sit;
> Cannot rest for a little bit.

Amazingly, this song was sung in the American public school system in the Philippines after the Second World War, although its meaning seems to have been lost on those who chose it as an anthem, however well-intentioned they may have been.

Culture and Rice: The Special Case of Japan

The importance of rice to Japanese society has been studied extensively and will provide a window through which to view its symbolic value in detail.

The concept of group harmony, dependency and consensus are thought to spring from wet rice cultivation. Historically, families pooled their labour and skills. Wet rice cultivation is labour-intensive, involving skills that everyone must execute simultaneously. This includes planting seedlings, building canals and dykes and sharing water resources, all of which linked families in an area. Houses were clustered together and the whole community helped out and did the planting together, for each household. The same was true at harvest time. Communal decisions and group interests were emphasized over individual preference. Efforts to avoid friction between families who would be neighbours and workmates for generations were paramount. This historic commitment to group synchronization, a feature of the original culture of rice, continues today and shapes group consciousness. Even though a small number of people grow rice in Japan today, 124 million people still try to sustain group harmony daily, and in confined spaces.

The Japanese language provides clues to these concepts

and values. The primacy of rice as a diet staple is echoed in the language. *Gohan* is both the word for 'cooked rice' and 'meal'. Adding prefixes to *gohan* gives us the words for breakfast (*asagohan*), lunch (*hirugohan*) and dinner (*bangohan*). These linguistic signals make it clear that thinking of a meal without rice is impossible.

Another indicator is the linguistic link to the early indigenous name of Japan, *mizuho no kuni* (the land of the water stalk plant or rice). Interestingly, the Japanese have identified the U.S. as *beikoku* (land of rice), thereby implying abundance.

Historically, rice has many links to Japanese culture. For example, the Emperor became a 'priest-king' early in Japanese history. His functions in Shintoism revolved around rice growing, and included making sake (rice wine) and *mochi* (rice cakes). Emperor Hirohito tended a rice plot on the Imperial grounds in Tokyo, until he became seriously ill, and even then, he worried about the weather and the crops. Traditions are maintained as the current Emperor Akihito blesses the rice crop. Many coronation ceremonies involving rice and rice products underscore links to the Emperor and to Shintoism.

Rice needed to be guarded, and represented security and prosperity, affirming its societal importance. *Sho*, a measure of rice, was used to determine wealth, in addition to being used as an instrument of trade, hard currency and as payment to samurai.

This small survey could be amplified by many examples

from other aspects of Japanese life, including folklore, festivals, art and family rituals. All parts of the rice plant are fully utilized. Every year some 32 kg of stalks are recycled into each tatami mat, which are used for flooring in many Japanese homes. Bran is made into face scrubs. Rice paste was used in bookbinding, a resist-dye technique for fabrics and especially in silk for kimonos. Rice is so enmeshed in the culture that, while people in the U.S. refer to 'the man in the moon' or see a woman's face, the Japanese see 'a rabbit pounding rice cakes' (*mochi*), a reminder of a popular folktale.

Recipes

Recipes and cooking methods reflect a time and place in history. Most recipes were transmitted orally until 'receipts' began to be written down, first for the professional chefs of aristocratic households and the clergy (whether at the time of the Roman Apicius or the early twentieth-century Escoffier), and later for the middle classes. To date, pre-modern Arabic culture produced the greatest number of cookbooks of which examples still exist. The first recipes were prose texts, which assumed highly skilled culinary competencies.

I have listed first examples of two recipes—pilaf and gumbo—that are popular and have variants in many countries, followed by an idiosyncratic mix of recipes that have wide appeal across cultures.

Aruzz Mufalfal

from Charles Perry's translation of a thirteenth-century recipe from a medieval Arabic cookery book called *Kitāb al Tabīkh*.

The way to make it is to take fat meat and cut it up medium. Melt fresh tail fat and throw away its cracklings, then throw the meat on it and stir until it is browned. Sprinkle a little salt and finely ground dry coriander on it. Then leave water to cover on it and boil it until it is done, and throw its scum away. Remove it from the pot after its liquid has dried up and it has started to stew, lest it be dry. Throw on as much dry coriander, cumin, cinnamon and finely ground mastic as it will bear, and likewise as much salt. When it is completely done, take it up from the pot, having been dried of moisture and fat. Sprinkle a little of those mentioned spices on it. Then take a measure of rice and three measures [and a half] of water. Melt fresh tail fat weighing one third as much as the meat. Throw the water in the pot. When it comes to [a boil], throw the melted fat on it. Throw mastic and sticks of cinnamon in it, then boil it until it comes to a full boil. Wash rice several times and colour it with saffron and throw it in the water; do not stir it. Then cover the pot awhile until the rice boils up and the water is boiling. Then open it and arrange that meat on top of the rice, and cover it with a cloth over the lid, and wrap it so that the air does not enter it. Then leave the pot until it grows quiet on a gentle fire for awhile, then take it up. Some people make it plain, not coloured with saffron.

Charles Perry's footnote indicates that 'aruzz' means rice, and 'mulfalfal' 'cooked to resemble peppercorns'; that is, as separate grains. He also mentions the possible influence of the Persian word '*pulau*' or 'pilaf'.

Pilaf with Golden Raisins and Pine Nuts

Pilaf can be made with bulgur and barley, but rice has pride of place. According to Claudia Roden in *The New Book of Middle Eastern Food* (2000), rice was introduced to Persia via India, and spread by Arabs southwest to Spain and as far south as Sicily. Called *roz* by Arabs, *pilav* by Turks, *chelow* by Iranians when plain and *polow* when other ingredients are added, pilaf can be served with stews, moulded, coloured red or yellow and otherwise prepared with vegetables, fruits, nuts, meat, fish, cream and milk. It can be served with all the other dishes or in sequence depending on where you are and with whom you are eating. Not surprisingly, there are many different kinds of long-grain rice used, each having its adherents.

My recipe for pilaf is a variation on Roden's classic Ottoman Empire Court recipe.

2 medium onions, chopped

15 ml (2 tbsp) canola oil

100 g (⅔ cup) pine nuts, toasted

400 g (2 cups) long-grain rice

675 ml (3 cups) chicken stock

1 tsp ground allspice

1 tsp cinnamon

1 tsp fenugreek seeds

salt and pepper

3 tbsp golden raisins

6 tbsp butter, cut into pieces

1 tbsp chopped dill

In a large pan, cook the onions in the oil until golden and soft. Add the pine nuts and rice and stir over a moderate heat until the rice grains and pine nuts are well coated with oil and starting to toast. Add the stock and stir in the allspice, cinnamon, fenugreek, salt, pepper and golden raisins. Bring to a boil, then simmer, covered, over a low heat for 20 minutes, or until the rice is tender. Stir in the butter and dill and serve hot.

Roden notes that for a Turkish variation you might add sautéed seasoned chicken livers and chopped dill to the hot rice.

Gumbo

Gumbo is a word, a recipe, a thick soupy stew, and a form of social commentary. The diverse ingredients that make up a gumbo could be said to represent a culinary dateline, where several strands of varying ethnic influences intersect, resulting in delicious and always interesting mixtures. Here are two versions, both from *The Carolina Rice Kitchen: The African Connection* (1992), with an Introduction to the annotated facsimile of *The Carolina Rice Cook Book*, compiled by Mrs Samuel G. Stoney in 1901.

New Orleans Gumbo

Take a turkey or fowl, cut it up with a piece of fresh beef, put them in a pot with a little lard, an onion and water sufficient to cook the meat. After they have become soft, add 100 oysters, with their liquor. Season to your taste, and just before taking up the soup, stir in, until it becomes mucilaginous, 2 spoonsful of pulverized sassafras leaves.

The phrase 'serve with rice' is missing from the text, but is assumed. Okra is also missing, but the sassafras presumably serves a similar thickening purpose.

Southern Gumbo

Slice 2 large onions, fry, have ready a good-sized chicken cut up; put in with the onions and fry brown. Have a quart of sliced okra and 4 large tomatoes; put all with the chicken in a stew pan and pour hot water over it. Let boil until thick; season with salt and red pepper pods. It must be dished and eaten with rice boiled.

Rice Puddings

Rice puddings exist in all rice cultures, both old and new. They vary in form from the simplest—soft and sweetened rice pudding—to *arroz con leche* made with condensed milk, or *riz a l'imperatrice*, a moulded creation of Escoffier's made in honour of the Empress Eugenie upon her marriage to Napoleon, which included vanilla custard, whipped cream and brandied fruit. In between are soft, firm, sliceable, spoonable, white, black and brown versions of rice pudding. Flavourings vary equally from vanilla and lemon zest, to cardamom, cashews, pistachios and saffron. Today rice puddings are made with dairy milk, soya milk, rice milk and coconut milk.

An American Colonial Rice Pudding

from J. M. Sanderson, *The Complete* Cook (1846)

Rice Custard.—Take a cup of whole Carolina rice, and seven cups of milk; boil it, by placing the pan in water, which must never be allowed to go off the boil until it thickens; then sweeten it, and add an ounce of sweet almonds pounded.

Kheer

2 pt (1.1 litres) whole milk

2 tbsp long-grain rice, such as basmati

4 whole green cardamom pods, lightly crushed

10 unsalted pistachios

2 tbsp sugar

To decorate

vark (edible silver or gold leaf, available from specialist

cake shops or some Asian grocers),

chopped pistachios, optional

Pour the milk into a heavy-based pan and heat gently (you can preheat the milk in a jug in the microwave, then transfer the hot milk to the pan, to save time, if you prefer). Add the rice and cardamom pods to the milk.

Slowly bring to the boil, then lower the heat and simmer rapidly, stirring from time to time to prevent the rice from sticking to the bottom of the pan. Simmer, stirring occasionally, until the milk is reduced by about half; this may take as long as 1¼ hours. While the milk is simmering, roughly chop the pistachios.

When the milk has reduced by half or more, remove and discard the cardamom pods. Transfer the rice pudding to a bowl. Add the sugar and taste, adding more sugar if you want it sweeter. Add the chopped pistachios, stir well, and leave to cool.

Cover the bowl with cling film and cool in the fridge for at least four hours or overnight.

When ready to serve, spoon into individual serving bowls. Decorate with vark, if using.

Sprinkle a few more chopped pistachios on top, if liked.

Sticky Rice Pudding
(for those who are lactose intolerant)

675 ml (3 cups) vanilla-flavoured rice milk

200 g (1 cup) sticky rice

150 g (¾ cup) granulated sugar

1 tbsp fresh ginger, grated

grated zest of 1 lemon

¼ tsp salt

1 vanilla bean pod, cut and scraped (save the pod for the

sugar container or vodka bottle)

2 large eggs

1 egg yolk

1 tbsp candied lemon peel, finely chopped

2 tbsp dark rum

Combine 2 cups of rice milk, rice, ½ cup of the sugar, the ginger and zest in a medium saucepan. Bring to a boil. Immediately reduce the heat to a simmer and cover. Cook until most of the milk is absorbed into the rice, about 20–25 minutes. Remove from heat. Uncover and cool, about 30 minutes.

Whisk together the remaining milk, the remaining sugar, salt, vanilla bean paste, the eggs and yolk in a medium bowl. When well mixed, strain the mixture through a sieve into a large saucepan. Cook, over low—medium heat, stirring constantly until the mixture coats the back of a spoon, about 8 minutes. Remove from heat.

Add candied lemon peel, rum and the cooled rice mixture, stirring until all is well blended. The mixture should be somewhat loose.

Transfer the pudding to a serving bowl or divide evenly among 6 oiled (almond oil, or very good olive oil) custard cups, which can be flipped over to remove the rice puddings once they are set. Chill for several hours before unmoulding. Serve the pudding after it has been at room temperature for one hour.

Soft coconut or mango sorbet is a good accompaniment, along with a few drops of dense, aged balsamic vinegar sprinkled on top.

Serves 6

Fried Rice

Fried rice is a preparation using leftover rice. Fried rice is probably far more common than any other way of using leftover cooked rice because it is fast and easy to prepare and uses up other leftovers in addition to the rice itself. Leftover rice has its own unique qualities and is prepared in ways that showcase them. All rice-consuming cultures have ways of using leftover rice that have become part of their culinary heritages.

Nasi Goreng

This recipe is the Indonesian and Malay variant of a Chinese way to make fried rice. I have adapted it from Michael Freeman, *Ricelands: The World of Southeast Asian Food* (London, 2008).

3 tbsp vegetable oil

3 cloves garlic, chopped

4 shallots, chopped

150 g (about 5 oz) raw prawns, peeled

150 g (about 5 oz) chicken cut into 5 cm (2 inches) pieces

1–2 tbsp light soy sauce

400 g (about 14 oz) leftover cold, cooked rice

4 eggs

2 spring onions, sliced to include some of the green stalks

3 medium-length fresh chillies, de-seeded and chopped

1 tbsp parsley, chopped

3 stalks coriander, leaves torn and chopped

pinch salt

pinch ground pepper

In a wok, heat the oil until almost smoking, then add the garlic and cook over a medium heat until it begins to turn golden-brown. Add the shallots and stir-fry until they begin to brown. Add the prawns, chicken and soy sauce and stir-fry until the prawns turn pink and the chicken loses its pinkness. Add the rice and stir continuously, mixing thoroughly with the prawns and chicken, for a few minutes, until hot. Cover and set aside. Fry the eggs in oil without breaking the yolks. Remove and set aside.

Scoop the rice out onto individual dishes, sprinkle the spring onions, chillies, parsley and coriander on top, add salt and pepper and finally place a fried egg on top of each.

Hoppin' John

Hoppin' John is an African-derived combination of rice and beans/pigeon peas whose history spans hundreds if not thousands of years, even though the recipe here is in modern format, and is made with modern rice. African and Asian rice, and all kinds of beans, are combined to create national dishes that have become part of the culinary profiles of different countries. Scholars of African and African-American foodways, including Karen Hess, Jessica Harris and James McWilliams, have speculated upon the origins of the term 'hoppin' John'. There is no consensus. *Arroz con frijoles* is the Latin version of rice and beans or hoppin' John, found throughout the Caribbean, Mexico and South America.

This recipe is adapted from Jessica Harris, *The Welcome Table: African American Heritage Cooking* (New York, 1995).

1 pound (450 g) dried black-eyed peas (cowpeas)

½ pound (225 g) salt pork

1 quart (950 ml) water

1 sprig fresh thyme

salt and freshly ground black pepper, to taste

1½ cups (150 g) raw long-grain rice

3 cups (675 ml) hot water

Pick over the black-eyed peas to remove dirt and stones. Soak them in water to cover at least 4 hours or overnight. Fry the salt pork in a large heavy casserole to render the fat. When the salt pork is crisp, add the black-eyed peas and the quart of water, the thyme, salt and pepper, cover, and cook over low heat for 40 minutes. Adjust the seasonings and continue to cook until the peas are tender. Add the rice, cover with the 3 cups hot water, and simmer over low heat until all of the liquid has been absorbed and the rice is tender. Serve hot.

On New Year's Day, in some families, a dime is placed in the hoppin' John to ensure special good luck throughout the year for the person who gets it.

Oysters and Rice: A Gullah Recipe

This recipe is from the Low Country: the coastal Sea Islands off the coast of Georgia (in the U.S.) where the African-American inhabitants are known as 'Gullahs', the 'people who eat rice'. It is adapted from Sallie Ann Robinson and Gregory Wrenn Smith, *Gullah Home Cooking the Daufuskie Way* (2003).

4 strips bacon

1 tablespoon cooking oil

1 large onion, chopped

1 medium-green bell pepper, chopped

2 tablespoons flour

3 cups (675 ml) warm water

salt and black pepper to taste

2 cups (400 g) uncooked rice

1 quart (1 kg) shucked oysters, drained

Fry the bacon until crisp in a medium pot. Remove the

bacon, leaving the grease in the pot. Add the oil, onion and bell pepper, and stir-fry until the onion is clear. Remove the onion and bell pepper, leaving the oil and grease. Brown the flour in the oil and grease, and then return the bacon, onion and bell pepper to the pot. Add the water, season to taste with salt and pepper, bring to a boil, lower the heat, and simmer 15 minutes, stirring often, to form a thin gravy. Rinse and strain the rice several times and rinse the oysters, then add both to the pot. Combine thoroughly, cover, and simmer, stirring occasionally, 30 to 45 minutes. Serve as a meal, with vegetable side dishes.

Rice Waffles

adapted from Fannie Merritt Farmer, *The Boston Cooking School Cook Book* (1896)

1 ¾ cups (600 g) flour

4 teaspoons baking powder

⅔ cup (90 g) cold cooked rice

¼ teaspoon salt

1 ½ cups (340 ml) milk

1 tablespoon melted butter

2 tablespoons sugar

1 egg

Mix and sift dry ingredients; work in rice with tips of fingers; add milk, yolk of egg well beaten, butter, and white of egg beaten stiff. Cook same as Waffles.

Rice à la Roast

The following recipe is adapted from a 1971 booklet published by the Rice Council of America in Houston, Texas. *Man-pleasing Recipes* begins with the phrase 'No man likes the same thing every night!' Rice recipes are given for breakfast, lunch and dinner. Adding zest to your family meals is the stated goal. 'Recipes for the vegetable part of the plate or for the elegant dinner. Tasty – try them!'

1 cup (125 g) chopped green onions, with tops

½ cup (60 g) chopped green pepper

2 tablespoons butter or margarine

3 cups (400 g) hot cooked rice, cooked in beef broth

3 tablespoons chopped pimiento

salt and pepper

Sauté onions and green pepper in butter until tender crisp. Add rice and pimiento. Toss lightly. Adjust seasonings to taste. Serve with your favourite roast.

Serves 6

Steamed Rice II

adapted from Gloria Bley Miller, *The Thousand Recipe Chinese Cookbook* (New York, 1970)

Wash rice thoroughly. Place in a pan with plenty of water. Boil for 5 minutes, stirring a few times. Drain.

Place the rice in a bamboo steamer lined with cheesecloth. Pierce the rice several times with chopsticks or fork, making small holes to let the steam pass through.

Cover the pan and steam over a medium heat for 20 minutes.

Miller notes that the liquid drained in step 2 can be mixed with sugar and served separately as a thin *congee*.

Rosematta Rice

adapted from Jeffrey Alford and Naomi Duguid, *Seductions of Rice* (New York, 2003)

This unusual parboiled red rice from South India is only partially milled: part of the reddish bran layer is left intact. Despite being parboiled, the rice is rinsed to remove debris and scum that forms when first cooking. It has an *umami* or meaty aroma and the grains easily stay separated from each other.

2 cups (400 g) rice

3 cups (675 ml) water (2¼ cups or 500 ml if using a rice cooker)

Wash the rice thoroughly under running water. The water will run reddish brown at first. Place in a sieve to drain. Pick through the rice and discard any hard kernels or other irregularities. Place in a heavy medium pot or in a rice cooker and add the water. If using a pot, bring to a vigorous boil, stir briefly, and boil uncovered for 3 to 4 minutes. Stir again, cover, and lower the heat to medium-low. Simmer for 5 minutes, then reduce the heat to very low and cook, still covered, for another 12 to 15 minutes. Let stand for another 10 to 15 minutes, without lifting the lid, to steam, then stir gently with a wooden paddle. The rice should be firm, bouncy even, and cooked through.

If using a rice cooker, turn on, cover and let cook. When the cooker turns off, let stand, covered for 10 to 15 minutes, before stirring.

Chicken Congee

Breakfast for many, comfort food for all: *congee* is often eaten with salty pickles, soy sauce, salted peanuts, radishes, pickled ginger, preserved greens, sausage, salted fish and any other leftovers that are handy. This recipe is adapted from Hsiang Ju Lin and Tsuifeng Lin, *Chinese Gastronomy* (New York, 1977).

100 g (½ cup) river rice (short-grain or other high-starch rice)

6 cups (1.35 litres) chicken stock

1 chicken breast

½ level teaspoon salt

2 tablespoons water

Wash rice. Bring it to the boil in chicken stock; then reduce heat to very low and simmer for 2 hours. Meanwhile, skin and bone the chicken breast and slice it with the grain. Flatten the slices with a few whacks of the side of the cleaver. Add salt and water. When the *congee* is ready to serve, turn off the heat. Stir the chicken slices, then stir them into the *congee* and let it stand for 3 to 4 minutes. Serve in bowls.

Pulot Hitam (Malaysian Black Rice Porridge)
adapted from Charmaine Solomon, *Encyclopedia of Asian Food*
(Boston, 1998)

220 g (1 cup) black glutinous rice

1.5 litres (3 pt) water

60 g (2 oz) palm sugar

2 tablespoons granulated sugar

2 strips pandan leaf

6 dried longans

250 ml (8 fl oz) coconut cream

¼ teaspoon salt

Wash rice in several changes of water and drain. Put into a heavy saucepan with the measured water and bring to the boil. Cover and simmer for 30–40 minutes, stirring occasionally to make sure the rice doesn't stick to the bottom of the pan. Add palm sugar and granulated sugar, pandan leaf and dried longans. (If the longans are still in their shells, discard the shells.) If the porridge becomes too thick, add more hot water. Continue cooking until the rice grains have become very soft. Serve warm, with coconut cream to which the salt has been added.

Serves 6

Brown Rice Horchata

adapted from www.massaorganics.com

½ cup (100 g) sugar

1 bag (7 oz, 350 g) unsweetened coconut flakes (use less
sugar if you use sweetened coconut flakes)

¾ cup (150 g) brown rice, soaked overnight and drained

1 cup (135 g) blanched almonds, toasted

1 cinnamon stick

¼ cup (55 ml) vanilla rice milk

Put sugar and 5 tablespoons water into a small saucepan, cover and boil over medium heat, swirling the pan occasionally, until the sugar dissolves, 4 to 5 minutes. Transfer to a bowl and allow the syrup to cool.

Put the coconut and 1½ cups water into a blender and purée until smooth. Strain through a fine sieve into a bowl, pressing on solids with a rubber spatula to extract as much coconut milk as possible, and set aside.

Put the rice, almonds, cinnamon and 2 cups water into clean blender and purée until smooth. Strain mixture through a cheesecloth-lined sieve into a bowl, pressing on solids to extract as much liquid as possible, and then return strained mixture to a clean blender. Add ¾ cup of the coconut milk, the syrup, rice milk and 2 cups ice cubes to blender and purée until ice is well chopped and drink is frothy. Divide between 2 to 4 glasses and serve immediately.

Serves 2–4

Pearly Meat Balls

adapted from Hsiang Ju Lin and Tsuifeng Lin, *Chinese Gastronomy*

(New York, 1977)

2½ oz (6 level tablespoons) glutinous rice

1 level teaspoon salt

¼ lb (100 g) fat pork, minced (ground)

¼ lb (100 g) lean pork, minced (ground)

1½ teaspoons wine

½ level teaspoon sugar

2 teaspoons light soy sauce

1 level tablespoon cornflour (cornstarch)

½ level teaspoon MSG

oil

5 tablespoons soy sauce

3 tablespoons vinegar

Put the rice in a 2-pint measuring cup (4-cup Pyrex bowl), and fill it to the 24 fluid oz (3 cup) mark with water. Let stand for 45 minutes, then drain it. Mix rice with salt.

In a separate bowl, mix together the pork, wine, sugar, soy sauce, cornflour, MSG and 1 tablespoon water. Blend the meat thoroughly with seasoning and shape it into 1-inch balls. Roll the balls in the glutinous rice and place them on an oiled plate of dish. Cover it closely and steam it for 30 minutes. Serve with soy sauce and vinegar, mixed together in a separate dish.

Epasol (Sweet Rice Flour Cakes)

adapted from Reynaldo Alejandro, *The Philippine Cookbook*

(New York, 1985)

This Philippine recipe comes with the admonition that Philippine etiquette involves serving all dishes or courses at once, including sweet and savoury dishes together. This is not necessarily a dessert.

4 cups (560 g) sweet rice flour

1½ cups (150 g) sugar

2 cans coconut milk

½ tsp salt

Toast the sweet rice flour on a cookie sheet. Bring sugar, coconut milk and salt to a boil. Add 3 cups toasted sweet rice flour. Mix well and cook until thick, stirring constantly.

Remove from heat and transfer to board well sprinkled with some of the reserved sweet rice flour. With a rolling pin, flatten to about ¼ inch and cut into diamonds. Roll in remaining rice flour.

Makes 15 to 20 cakes

Rice Noodles

Rice noodles can be fresh or dried, long or short, thin or thick, and may even come in pre-cooked broad sheets that can be cut or shaped as you wish. China and Thailand are the largest exporters of rice noodles. Some are made with rice flour, others with sticky rice flour, and yet others with different flours (such as tapioca or cornstarch) for springiness and texture.

Fresh rice sheets can be sliced or stuffed, and must be eaten before they dry out. Dried rice noodles, also called rice sticks, should be soaked in room- temperature water for 10 to 20 minutes to soften, and then added to the cooking process, otherwise they will become mushy. To remove excess starch, so as to keep the broth clear if you are making soup, rice noodles can be briefly boiled after soaking.

Singapore Noodles

adapted from Corinne Trang, *Noodles Every Day: Delicious Asian Recipes from Ramen to Rice Sticks* (San Francisco, 2009)

This recipe is popular in Chinese restaurants in the U.S., and uses leftover Cantonese roast pork and curry powder, which showcases Indian influence in Singapore.

8 oz (225 g) dried rice vermicelli, soaked in water until pliable

24 small tiger shrimp, heads removed, peeled and deveined

3 tbsp vegetable oil

1 small onion, cut into small wedges

½ cup (75 g) fresh shelled peas, or thawed frozen peas

2 tsp curry powder (Indian)

6 oz (175 g) Cantonese roast pork

1½ tbsp fish sauce

Kosher salt and freshly ground pepper

6 sprigs coriander, trimmed

Cook the noodles in a pot of boiling water until tender, about 10 seconds. Transfer them to a bowl. In same water cook the shrimps, about 1 minute.

Heat 1 tablespoon of oil in a skillet or wok, and stir-fry onion until golden, about 3–4 minutes. Add other 2 tablespoons oil, the noodles and peas. Sprinkle curry powder on top. Toss well, making sure all noodles become yellow. Add pork, shrimp, fish sauce and heat thoroughly, stir frying all the while, about 5 minutes. Adjust seasoning with salt and pepper, and serve with coriander.

Serves 6

Select Bibliography

Achaya, K. T., *Historical Dictionary of Indian Food* (Oxford, 2001)

Al-Baghdadi, Mohammad Ibn Al-Hasah, *A Baghdad* Cookery Book (Blackawton, Totnes, Devon, 2006)

Alcock, Joan P., *Food in the Ancient World* (London, 2006)

Anderson, E. N., *The Food of China* (New Haven, CT, 1990)

Balfour, Edward, *The Cyclopaedia of India and of Eastern and Southern Asia: Commercial, Industrial and Scientific Products of the Mineral, Animal and Vegetable Kingdoms, Useful Arts and Manufacture* (London, 1885)

Barnes, Cynthia, 'The Art of Rice', *Humanities*, 24 (September–October 2003), www.neh.gov

Beeton, Isabella Mary, ed., *Mrs Beeton's Book of Household Management* (London, 1861)

Boesch, Mark J., *The World of Rice: Its History, Geography and Science* (New York, 1967)

Bray, Francesca, *The Rice Economies: Technology and Development in Asian Societies* (Berkeley, CA, 1994)

Burton, David, T*he Raj at Table: A Culinary History of the British in India* (London, 1993)

Carney, Judith, 'The African Antecedents of Uncle Ben in

U.S. Rice History', *Journal of Historical Geography*, 29 (1 January 2003), pp. 1–21

—, 'African Rice in the Columbian Exchange', *Journal of African History*, XLII/3 (2001), pp. 377–96

—, 'From Hands to Tutors: African Expertise in the South Carolina Rice Economy', *Agricultural History*, 67 (Summer 1993), pp. 1–30

—, '"With Grains in Her Hair": Rice in Colonial Brazil', *Slavery and Abolition*, XXV/1 (2004), pp. 1–27

Coclanis, Peter A., 'The Poetics of American Agriculture: The United States Rice Industry in International Perspective', *Agricultural History*, 69 (Spring 1995), pp. 140–62

—, *The Shadow of a Dream: Economic Life and Death in the South Carolina Low Country, 1670–1920* (New York, 1989)

—, 'Southeast Asia's Incorporation into the World Rice Market: A Revisionist View', *Journal of Southeast Asian Studies*, XXIV/2 (1993), pp. 251–67

—, 'Breaking New Ground: From the History of Agriculture to the History of Food Systems', *Historical Methods*, 38 (Winter 2005), pp. 5–15

Cole, Arthur Harrison, *Wholesale Commodity Prices in the United States, 1700–1861* (Cambridge, MA, 1938)

Collingham, Lizzie, *Curry: A Tale of Cooks and Conquerors* (Oxford and New York, 2006)

Corson, Trevor, *The Story of Sushi: An Unlikely Saga of Raw Fish and Rice* (New York, 2008)

Davidson, Alan, *The Oxford Companion to Food,* 2nd edn

(New York, 2006)

Davis, Lucille, *Court Dishes of China: The Cuisines of the Ch'ing Dynasty* (Rutland, VT, and Tokyo, 1966)

Dethloff, Henry C., *A History of the American Rice Industry, 1685–1985* (College Station, TX, 1988)

Ewing, J. C., *Creole Mammy Rice Recipes* (Crowley, LA, 1921)

Fragner, Bert, From the Caucasus to the Roof of the World: A Culinary Adventure', in Sami Zubaida and Richard Tapper, eds, *Culinary Cultures of the Middle East* (London, 1994)

Freeman, Michael, *Ricelands: The World of South-East Asian Food* (London, 2008)

Grist, D. H., *Rice*, 6th edn (New York, 1986)

Hall, Gwendolyn Midlo, *Africans in Colonial Louisiana: The Development of Afro-Creole Culture in the Eighteenth Century* (Baton Rouge, LA, 1992)

Hansen, Eric, 'The Nonya Cuisine of Malaysia: Fragrant Feasts Where the Trade Winds Meet', *Saudi Aramco World*, 54 (September–October 2003), pp. 32– 9

Harris, Jessica, *Beyond Gumbo: Creole Fusion Food from the Atlantic Rim* (New York, 2003)

—, *Iron Pots and Wooden Spoons: Africa's Gifts to New World Cooking* (New York, 1989)

Hess, Karen, *The Carolina Rice Kitchen: The African Connection* (Columbia, SC, 1992)

Higham, Charles, and Tracey L.-D Lu, 'The Origins and Dispersal of Rice Cultivation', *Antiquity*, 72 (December 1998), pp. 867–77

Huggan, Robert D., 'Co-Evolution of Rice and Humans', *GeoJournal*, 35 (1995), pp. 262–5

Kingston, Maxine Hong, *The Woman Warriors: Memoirs of a Girlhood Among Ghosts* (New York, 1975)

Kumar, Tuk-Tuk, *History of Rice in India: Mythology, Culture and Agriculture* (New Delhi, 1988)

Latham, A.J.H., *Rice: The Primary Commodity* (London and New York, 1998)

McWilliams, James E., *A Revolution in Eating: How the Quest for Food Shaped America* (New York, 2005)

Mancall, Peter C., Joshua L. Rosenbloom and Thomas Weiss, 'Slave Prices and the Economy of the Lower South, 1722–1809', conference paper, January 2000, at www. eh.net.

Medina, F. Xavier, *Food Culture in Spain* (Westport, CT, 2005)

Mintz, Sidney W., 'Asia's Contributions to World Cuisine', in Sidney C. H. Cheung and Tan Chee-Beng, eds, *Food and Foodways in Asia* (Abingdon, 2007), pp. 201–10

Ohnuki-Tierney, Emiko, 'Rice as Self: Japanese Identities Through Time', *Education About Asia*, 9 (Winter 2004), pp. 4–9

Owen, Sri, *The Rice Book* (London, 1993)

Piper, Jacqueline M., *Rice in South-East Asia: Cultures and Landscapes* (New York, 1994)

Robinson, Sallie Ann, with Gregory Wrenn Smith, *Gullah Home Cooking the Daufuskie Way: Smokin' Joe Butter Beans, O' 'Fuskie Fried Rice, Sticky- Bush Blackberry Dumpling, and Other Sea Island Favorites* (Chapel Hill, NC, 2003)

Roden, Claudia, *Arabesque. A Taste of Morocco, Turkey, and Lebanon* (New York, 2006)

—, *The Food of Spain* (New York, 2011).

—, *The New Book of Middle Eastern Food*, revd edn (New York, 2000)

Rodinson, Maxime, A. J. Arberry and Charles Perry, *Medieval Arab Cookery: Essays and Translations* (Blackawton, Totnes, Devon, 2001)

Sen, Colleen Taylor, *Curry: A Global History* (London, 2009)

Simmons, Marie, *The Amazing World of Rice: With 150 Recipes for Pilafs, Paellas, Puddings, and More* (New York, 2002)

Smith, Andrew, ed., *The Oxford Encylopedia of Food and Drink in America* (Oxford, 2004)

Smith, C. Wayne, and Robert Henry Dilday, eds, *Rice: Origins, History, Technology, and Production* (Hoboken, NJ, 2003)

Sokolov, Raymond, 'A Matter of Taste: A Two-Faced Grain', *Natural History*, 102 (January 1993), pp. 68–70

Walker, Harlan, *Staple Foods: Proceedings of the Oxford Symposium on Food and Cookery* (Blackawton, Totnes, Devon, 1990)

West, Jean M., 'Rice and Slavery: A Fatal Gold Seede', www.slaveryinamerica.org, accessed 22 April 2011

Wright, Clifford, *A Mediterranean Feast: The Story of the Birth of the Celebrated Cuisines of the Mediterranean, from the Merchants of Venice to the Barbary Corsairs* (New York, 1999)

Yin-Fei Lo, Eileen, *The Chinese Kitchen: Recipes, Techniques, History, and Memories from America's Leading Authority on Chinese Cooking* (New York, 1999)

Zafaralla, P. B., *Rice in the Seven Arts* (Laguna, Philippines, 2004)

Zaouali, Lilia, M. B. DeBevoise and Charles Perry, *Medieval Cuisine of the Islamic World: A Concise History with 174 Recipes* (Berkeley, CA, 2009)

Zubaida, Sami, and Richard Tapper, eds, *A Taste of Thyme: Culinary Cultures of the Middle East* (London and New York, 2000)

Websites and Associations

Rice Production and Research

Asia Rice
http://asiarice.org

Western Farm Press
http://westernfarmpress.com

American Association of Cereal Chemists
www.aaccnet.org

Africa Rice
www.africarice.org

Asia Society
www.asiasociety.org

The Chartered Institute of Architectural Technologists
www.ciat.cgiar.org

Economic Research Service, USDA
www.ers.usda.gov

Euromonitor

www.euromonitor.com

Food and Agriculture Organization

www.fao.gov

Food Timeline

www.foodtimeline.org

Grain

www.grain.org

International Rice Research Institute

www.irri.org

Lotus Foods

www.lotusfoods.com

Riceweb

www.riceweb.org

U.S. Rice Producers

www.usriceproducers.com

Recipes

Flavor and Fortune

www.flavorandfortune.com

Clifford A. Wright

www.cliffordawright.com

Mex Connect

www.mexconnect.com

On the Table

www.onthetable.us

Sri Owen

www.sriowen.com

Acknowledgements

I would like to thank the following individuals for their help while I researched *Rice: A Global History*: Jay Barksdale, Phil Bruno, Robert Carmack, Amy Cole, Dori Erlich, Barry Estabrook, Suzanne Fass, Alex Garcia, Joan Giurdanella, Jenny Huston, J. J. Jacobson, Rachel Laudan, Monique Lignon, Charlotte Lindberg, Mai Ling, Jan Longone, Vanessa Lucin, Danielle Marton, Gary Marton, Simone Marton, Joanna McNamara, Hung Nguyen, Margaret Happel Perry, Morrison Polkinghorne, Judy Rusignolo, Marie Simmons, Andy Smith, Jane Stanicki, Rick Stein, Gary Taubes, Laura Weiss, David Wexler, Stacia Wilkie, Sarah Wormer and especially Ed Smith.

I would also like to thank the New York Public Library, the Clements Library in Ann Arbor, Michigan, and libraries in cities everywhere, without which our lives would be inestimably poorer. Databases, while invaluable, are not enough.

While I received much help from many people, any errors are mine.

Photo Acknowledgements

The author and publishers wish to express their thanks to the below sources of illustrative material and/or permission to reproduce it. Some locations of artworks are also given below.

Courtesy http://allhindugods.blogspot.com/2013/01/pongalkolam.html: p. 95; Asian Art Museum, Toronto: p. 94; photos by the author: pp. 10, 23, 25, 51; photo bdspn/iStockphoto: p. 60 (foot); from 'Mrs Beeton', *The Book of Household Management. . . by Isabella Mary Beeton* (London, 1861): p. 43; bonchan/iStockphoto: p. 13; bopav/iStockphoto: p. 89; British Library, London (photos © The British Library Board): pp. 33, 96; photos © The Trustees of the British Museum, London: pp. 59; photos Amy Cole: pp. 9, 85; graytown/iStockphoto: p. 49; photo Hargrett Rare Book and Manuscript Library/University of Georgia Libraries: p. 44; harikarn/iStockphoto: p. 76; photo ildi/iStockphoto: p. 54; photo Jastrow: p. 92; Library of Congress, Washington, DC: pp. 46, 72; lilly3/iStockphoto: p. 57; MickyWiswedel/iStockphoto: p. 4; © Rachel Park from art@potalaworld.com: p. 22; piotr_malczyk/iStockphoto: p. 5; Quaker Puffed Rice Machine–image courtesy of the Anderson Center: p.

71; photo quintanilla (© CanStockphoto Inc., 2013): p. 35; robynmac/iStockphoto: p. 73; photo Jon Sullivan: p. 48; from William Tayler, *Sketches illustrating the manner and customs of the Indians and the Anglo-Indians* (London, 1842): p. 59; typssiaod/iStockphoto: p. 14; photo courtesy U.S. Department of Agriculture: p. 20; USA Rice Federation: p. 64; wagner_christian/iStockphoto: p. 24.

图书在版编目（CIP）数据

大米 ／（美）勒妮·马顿著；王艺蒨译．——
北京：北京联合出版公司，2024.4
（食物小传）
ISBN 978-7-5596-7361-9

Ⅰ．①大…　Ⅱ．①勒…　②王…　Ⅲ．①大米－饮食－
文化－世界－普及读物　Ⅳ．① TS972.131-49

中国国家版本馆 CIP 数据核字（2024）第 022627 号

大米

作　　者：〔美国〕勒妮·马顿
译　　者：王艺蒨
出 品 人：赵红仕
责任编辑：夏应鹏
产品经理：汤　成　徐　静
装帧设计：鹏飞艺术
封面插画：〔印度尼西亚〕亚尼·哈姆迪

北京联合出版公司出版
（北京市西城区德外大街 83 号楼 9 层　　100088）
北京天恒嘉业印刷有限公司印刷　　新华书店经销
字数 124 千字　889 毫米 ×1194 毫米　1/32　8.75 印张
2024 年 4 月第 1 版　　2024 年 4 月第 1 次印刷
ISBN 978-7-5596-7361-9
定价：59.80 元